Principles of
Mobile Computing
and Communications

Principles of
Mobile Computing
and Communications

Mazliza Othman

CRC Press
Taylor & Francis Group
Boca Raton London New York

CRC Press is an imprint of the
Taylor & Francis Group, an **informa** business

CRC Press
Taylor & Francis Group
6000 Broken Sound Parkway NW, Suite 300
Boca Raton, FL 33487-2742

First issued in paperback 2019

© 2008 by Taylor & Francis Group, LLC
CRC Press is an imprint of Taylor & Francis Group, an Informa business

No claim to original U.S. Government works

ISBN-13: 978-1-4200-6158-1 (hbk)
ISBN-13: 978-0-367-38814-0 (pbk)

Library of Congress Cataloging-in-Publication Data

Othman, Mazliza.
  Principles of mobile computing and communications / Mazliza Othman.
    p. cm.
  Includes bibliographical references and index.
  ISBN 978-1-4200-6158-1 (alk. paper)
  1. Mobile communication systems. 2. Mobile computing. I. Title.

TK6570.M6O84 2007
621.3845'6--dc22   2007012440

**Visit the Taylor & Francis Web site at
http://www.taylorandfrancis.com**

**and the CRC Press Web site at
http://www.crcpress.com**

Contents

Preface

This book is written to address a number of issues that currently are not addressed by other books on the topics of mobile computing and communications. I have taught a mobile computing course for a few years and have come across a number of books that discuss wireless network technologies and infrastructures, and books that focus on tools and software to develop mobile applications. What I find missing in these books is a discussion on how developing mobile computing applications are different from developing conventional applications, the issues and constraints that need to be addressed, and why mobile applications are different from conventional applications. This book is my attempt at addressing those shortcomings.

Another difficulty that I encountered when teaching this subject is that most books on wireless networks are written for engineering students. Adopting the material for computer science students is quite a task. That is another reason for this book—it is written specifically for computer science students and people from a computer or information technology background.

Overview of the Book

This book can be used as a textbook for a mobile computing course (introductory or intermediate). It is targeted at second- or third-year undergraduate computer science students, but can also be used as a reference book for a postgraduate course. It assumes that readers have a basic knowledge of computer communication networks. If enough time and motivation exist, the reader is advised to go through the entire book cover to cover. Otherwise, the reader may choose topics of interest. The book is written so that the chapters are independent of each other. It is organized as follows.

Chapter 1 gives an overview of what mobile computing has to offer and how mobile applications will eventually change the way we work and live. It describes Mark Weiser's vision of a ubiquitous computing environment and proceeds to give examples of mobile applications in different fields followed by a section that gives an overview of the evolution of wireless networks and services.

The next five chapters discuss the underlying network technologies required to support such applications.

Chapters 2–6 focus on the technologies and infrastructure of the following wireless networks: cellular networks, wireless local area networks, personal area networks, sensor networks, and mobile ad hoc networks. Each chapter discusses the relevant standards and services supported by the standard. The security issues related to each network are also explored. Chapter 2 gives a fundamental understanding of the wireless network infrastructure and protocols. Chapters 3–6 can be read independently of each other without affecting understanding of later chapters.

Chapter 7 explains why existing Internet protocols are unsuitable for mobility support and proceeds to discuss the Mobile IP standard that was designed to support roaming users. In addition to registration messages, routing, and tunneling, this chapter also discusses how security is addressed in Mobile IP by extending it to support accounting, authentication, and authorization services. This chapter is optional.

Chapter 8 is fundamental to understanding mobile computing issues. It discusses various issues and why these must be considered when developing mobile applications. The objective is to highlight the differences between developing desktop and mobile applications. Among issues presented are adaptive behavior, power management, resource constraints, interface design, and seamless mobility support. The Odyssey, Spectra, and Aura projects are among examples used to illustrate the complexity of designing and developing smart mobile applications.

Chapter 9 focuses on location-sensing techniques and systems. It explains why identifying a user's location is important to delivering context-sensitive information. After defining key terms, it discusses three location-sensing techniques followed by a brief taxonomy of location systems. The remaining sections present case studies of the implementation of location systems in a hospital and a tracking application.

Finally, Chapter 10 discusses security issues that have not been covered in previous chapters.

In each chapter, relevant case studies are used as examples so that the reader can better understand the practical aspects and how a problem is addressed. Reference, bibliography, and online resource lists are provided at the end of each chapter so that readers may further explore the topic. I hope readers will find this book interesting and useful.

Acronyms

The telecommunications field is famous for its notorious use of acronyms. A list of acronyms used in this book is provided to help minimize confusion.

Acknowledgment

I would like to express my gratitude to Nor Edzan Che Nasir for proofreading the initial drafts of the manuscript. Her comments have made it much more readable.

About the Author

Mazliza Othman graduated from Universiti Kebangsaan Malaysia with a B.Sc. in computer science, and she worked at a telecommunications company briefly before going to the United Kingdom to pursue her postgraduate studies. She obtained a M.Sc. in data communication networks and distributed systems and a Ph.D. from the University of London. Currently, she is in the Faculty of Computer Science and Information Technology, University of Malaya. Her main areas of interest are distributed systems and mobile computing. She has published papers and articles in these areas of research in international journals and conference proceedings.

About the Author

Chapter 1

Introduction

The most familiar aspect of mobile computing technology is the hand phone. About two decades ago, a hand phone was bulky and was only used for voice communication. It was merely an extension of the fixed line telephony that allowed users to keep in touch with colleagues. Now the hand phone is not only used for voice communication, it is also used to send text and multimedia messages. Future mobile devices will not only enable Internet access, but will also support high-speed data services.

In addition to the hand phone, various types of mobile devices are now available, for example, personal digital assistants (PDAs) and pocket personal computers (PCs). Road warriors use mobile devices to access up-to-date information from the corporate database. A police officer at a crime scene may send a fingerprint picked up there for matching with data in a central database through a wireless network, hence leading to faster identification and arrest of potential suspects. The global positioning system (GPS) is used in search and rescue missions, for monitoring and preservation of wildlife, and for vehicle theft prevention. Though many of us are unaware of when mobile computing technology is being used, it has permeated all aspects of our lives.

What is mobile computing? Simply defined, it is the use of a wireless network infrastructure to provide anytime, anywhere communications and access to information. There are many aspects of mobile computing and, sometimes, different terms are used to refer to them. This chapter gives an overview of what mobile computing has to offer and how it improves the quality of our lives. Later chapters discuss the underlying wireless networks and technologies that make mobile computing applications possible.

1.1   Mobile Computing Applications

In 1991, Mark Weiser envisioned the next-generation computer technologies that "weave themselves into the fabric of everyday life until they are indistinguishable from it." He described a ubiquitous computing environment that enhances the environment by making many computers available throughout the physical realm, while making them effectively invisible to the user. Weiser pointed out that anthropological studies of work life showed that people primarily work in a world of shared situations and unexamined technological skills. Today's computer technology does not conform to this description because it remains the focus of attention instead of being a tool that disappears from users' awareness. Ubiquitous computing aims to make computers widely available throughout users' environments and effortless to use. In other words, users should be able to work with computing devices without having to acquire the technological skills to use them. Computers are integrated into their environments so that users are not even aware that they are using a computer to accomplish a task. Unlike the computer technology of today, users need not acquire specific skills to use computers because their use would be intuitive. The aim of ubiquitous computing is to create a new relationship between people and computers in which the computers are kept out of the way of users as they go about their lives.

Instead of computers that sit passively on desks, ubiquitous computers are aware of their surroundings and locations. They come in different sizes, each tailored to a specific task. At the Xerox Lab, Weiser and his colleagues developed a tab that is analogous to a Post-it® note, a pad that is analogous to a sheet of paper, and a board that is analogous to a yard-scale display. An office may contain hundreds of tabs, tens of pads, and one or two boards. These devices are not personal computers, but are a pervasive part of everyday life, with users often having many units in simultaneous operation. Unlike a laptop or a notebook, which is associated with a particular user, tabs and pads can be grabbed and used anywhere—they have no individualized identity and importance. You may have a few pads on your desk, each dedicated to a particular task in the same way that you spread papers on your desk.

An employee ID card is replaced by an active badge that is of the same size. It identifies itself to receivers placed throughout a building. It makes it possible to keep track of people and objects that it is attached to. Because an active badge is associated to a particular user, it can become a form of ID; for example, an electronic door to a restricted area would only open to authorized users. When I am not in my office, the system detects my current location and forwards phone calls there. When I walk into a lecture hall, the system detects my presence, checks the timetable, deduces that I am there for a WRES3405 lecture, and automatically downloads that day's lecture notes. When the system detects the presence of a team working on a particular project in a meeting room, it checks the room booking system to determine if there is a scheduled meeting. Confirming that it is indeed the case, it downloads and displays the previous minutes on the board. As the meeting progresses, team members may manipulate the board using a tetherless pen that need not touch the screen, but can operate from a few meters away. Using the pen, a team member may point to an object on the board, select it, and modify it.

Pervasive computing is a term that is synonymous with ubiquitous computing. Many interesting projects on pervasive computing are carried out at Carnegie-Mellon University. The Portable Help Desk (PHD) is an application developed under Project Aura that makes use of spatial (a user's relative and absolute position and orientation) and temporal (scheduled time of private and public events) awareness. PHD allows a user to determine the location of colleagues and information about them. It is equipped with the capability to display maps of surrounding areas, indicating resources and nearby people. It also notifies users of the availability of resources they may need, for example, a nearby printer or café. PHD is equipped with visual and audio interfaces, each of which provides support in different contexts; for example, a user who is walking is more likely to prefer an audio interface to interact with PHD.

An important requirement of the applications discussed so far is that for them to offer relevant information to the user (e.g., a café is about 100 m to your left), they need to be aware of their context. This is a very important aspect of mobile computing applications. To be useful, an application needs to be aware of its current environment. For example, if I am currently in Kuching and I request information about seafood restaurants, I expect the application to give me a list of seafood restaurants in Kuching, not Kuala Lumpur. For this reason, a tourist guide application must have context awareness embedded so that it can deliver information that is relevant to the users.

HyperAudio and HIPS (Petrelli et al. 2001) are handheld electronic museum guides that adapt their behavior to that of a visitor. A visitor to a museum is given a handheld device equipped with headphones. As the visitor approaches an exhibit, the system dynamically composes a presentation of the object in sight. When the system detects that a visitor pauses in front of a display, it presents information about it. The information is presented in the form of audio recording, a relevant image, and a set of links for obtaining more information about it. The system obtains an estimate of the distance between the visitor and the display and adapts the way the information is presented. For example, if it detects that the visitor is standing right in front of the display, the audio message would say, "this item is. . . ." If the visitor is a distance away, it may attract his attention to it by saying, "the display in front of you . . ." or "the display to your left is. . . ." The system deduces that the visitor is very interested in the display if he or she pauses in front of it for more than a certain period of time and proceeds to present more detailed information about it. As the visitor moves away to view another display, the system detects the distance between the current display and the next one and starts to download the presentation for the next display.

Another class of information that makes use of wireless technology is wearable computing, which involves integrating computers into our clothes to perform certain functions, for example, monitor the wearer's heartbeat and blood pressure. There are many practical and useful applications of wearable computing. Guide dogs and canes are very useful in assisting visually impaired people to avoid obstacles and negotiate changes in ground level, such as steps. However, they are not helpful in avoiding higher obstacles such as street signs and tree branches. This difficulty may be overcome with the use of a wearable headset consisting of a laptop, a video camera with infrared (IR) light emitting diodes mounted on one side of an eyeglass frame, and a scanning fiber display and optics mounted in a tube. The software comprises a machine vision program that identifies potential collision objects, a program that controls the display, and a graphical user interface (GUI) to help set parameters for the embedded processors and generate bright warning icons.

A more recent technology is a wireless sensor network (WSN). In a WSN, sensors are placed at strategic locations to monitor certain aspects of the environment. For example, biologists may use it for habitat monitoring to study behavioral patterns of a species. The use of sensor networks assists ecologists to accurately measure the degree of microenvironmental variance that organisms experience (Szewczyk et al. 2004). Data collected by scientists regarding population dynamics and habitat needs is important in conservation biology, landscape monitoring and management, and species-recovery efforts. Sensor nodes are also used to monitor personnel and mobile assets; for example, an alarm is triggered when a printer is detected leaving an office area without authorization. One application of this technology is in agriculture, where sensors are used to monitor environmental conditions that may affect the crop. Early detection and alert of a change in temperature, for example, would help farmers to take precautionary steps to protect their crops.

Another novel invention using wireless technology is the virtual fence (Murray 2004). Cowboys on horsebacks herding cattle might one day become a feature of a bygone era as the introduction of virtual fences allows ranchers to herd their cattle from the comfort of their homes. The virtual fence is downloaded to the cows by transmitting GPS coordinates to head collars worn by the cows. The dynamic virtual fences are moved along desired trajectories. The collars are equipped with a wireless fidelity (Wi-Fi) networking card, a Zaurus PDA, an eTrex GPS unit, and a loudspeaker that transmits occurring sounds (e.g., roaring tigers, barking dogs) when a cow strays from the intended path. This multidisciplinary project, the brain child of a biologist, is made possible in collaboration with computer scientists.

Sensor technology can potentially play an important role in search and rescue operations by first responders (i.e., emergency personnel), such as firefighters, paramedics, and police, who arrive at the scene immediately after an event (e.g., a fire, an earthquake, a building collapse) occurs. Firefighters wear tags to allow easy tracking of their movements to coordinate search and rescue operations more effectively. The firefighters can be informed if a particular section of a building is found to be unstable and is about to collapse, and they are directed to evacuate it immediately. A wireless vital sign monitor is attached to victims found trapped so that their condition can be monitored to ensure that they receive the appropriate medical attention as soon as they are rescued. This noninvasive sensor monitors vital signs such as heart rate, oxygen saturation, and serum chemistry measurements. The vital sign monitor helps the paramedic team determine which victims' conditions are more critical so that they can prioritize medical attention to more severely injured victims. The application and architecture required to support this emergency response application is being developed under the CodeBlue project at Harvard University.

Wireless technology is also used in healthcare. The Arrhythmia Monitoring System (AMS) is a medical telemetry (telemedicine) system that makes use of wireless technology to monitor patients suffering from arrhythmia (Liszka et al. 2004). Among the complications that arise from arrhythmia are the loss of regular heartbeat and subsequent loss of function and rapid heartbeats. AMS provides a means for healthcare professionals to continuously monitor a patient's electrical cardiac rhythms remotely even though the patient is not at the hospital. This technology allows patients to be in the comfort of their homes without jeopardizing their health. It is also useful for monitoring the heart functions of astronauts who are more susceptible to cardiac dysrhythmias when in space.

The system architecture consists of a wearable server, a central server, and a call center. The wearable server is a small communications device worn by the patient that collects the patient's electrocardiogram ([ECG], i.e., the heart muscles' electrical activity). The data is collected using wires attached to skin-contact biosensors. The wearable server receives analog signals from the sensors and converts them into digital signals. Data is collected every 4 ms and requires a minimum baud rate of 22.5 kbps to transmit over a wireless link to the central server.

The central server is located close to the patient. Its functions are data compression, location awareness utilizing GPS, and rudimentary arrhythmia detection. It serves as a wireless gateway to a long-distance cellular network. Data is routed via the call center that is manned 24/7, by healthcare professionals who monitor the ECG signals and respond to alerts. The system transmits an alert automatically if it detects that the patient is about to have or is having an arrhythmia attack. A patient can press a button on the wearable server to send a noncritical alert to the call center if the heart flutters or other unusual feeling occurs. There is also a panic button that a patient can press to send a critical alert for help so that an emergency response team can be rushed to the most recent GPS location.

The GPS location service is a critical part of the system as it is imperative that an emergency response team is dispatched in the quickest time possible. A patient's location is tracked using a GPS transceiver equipped with a 1.55 GHz GPS antenna and a 2.4 GHz Bluetooth antenna. The location information is sent to the receiver every 10 s and acquires a minimum of three GSP satellite signals. A patient's location can be accurately tracked within 10 m.

Another category of mobile applications that is gaining popularity is mobile commerce or m-commerce, which is likely to become an important application of this technology. M-commerce application can be classified into 10 types (Varshney and Vetter 2002):

1. Mobile financial application (business-to-customer [B2C] and business-to-business [B2B]): The mobile device is used as a powerful financial medium.
2. Mobile advertising (B2C): It turns the wireless infrastructure and devices into a powerful marketing medium.
3. Mobile inventory management (B2C and B2B) or product locating and shopping (B2C and B2B): It is an attempt to reduce the amount of inventory needed by managing in-house and on-the-move inventory. It also includes applications that help to locate products and services that are needed.
4. Proactive service management (B2C and B2B): It attempts to locate products and services that are needed.
5. Wireless reengineering (B2C and B2B): It focuses on improving the quality of business services using mobile devices and wireless infrastructure.
6. Mobile auction or reverse auction (B2C and B2B): It allows users to buy or sell certain items using multicast support of wireless infrastructure.
7. Mobile entertainment services and games (B2C): It provides entertainment services to users on a per-event or subscription basis.
8. Mobile office (B2C): It provides the complete office environment to mobile users anywhere, anytime.
9. Mobile distance education (B2C): It extends distance or virtual education support for mobile uses everywhere.
10. Wireless data center (B2C and B2B): It supports large amounts of stored data to be made available to mobile users for making "intelligent" decisions.

The mobile computing applications discussed so far provide a glimpse of what mobile computing technology has to offer. The applications are used in many different fields and may perform generic functions or be tailored to specific needs. The next section gives an overview the evolution of wireless networks that have made these applications possible.

1.2   Evolution of Wireless Networks and Services

The first generation (1G) wireless network was analog. The first in North America was advanced mobile phone system (AMPS), which was based on frequency division multiple access. A total of 1664 channels were available in the 824 to 849 MHz and 869 to 894 MHz band, providing 832 downlink (DL) and 832 uplink (UL) channels. AMPS, widely used in North America, supports frequency reuse. The underlying network is a cellular network where a geographical region is divided into cells. A base station (BS) at the center of the cell transmits signals to and from users within the cell.

The second generation (2G) systems onward are digital. Digital systems make possible an array of new services such as caller ID. The Global System for Mobile Communications (GSM) is a popular 2G system. GSM offers a data rate of 9.6 to 14.4 kbps. It supports international roaming, which means users may have access to wireless services even when traveling abroad. The most popular service offered by GSM is the Short Message Service (SMS), which allows users to send text messages up to 160 characters long.

2.5G systems support more than just voice communications. In addition to text messaging, 2.5G systems offer a data rate on the order of 100 kbps to support various data technologies, such as Internet access. Most 2.5G systems implement packet switching. The 2.5G systems help provide seamless transition technology between 2G and third generation (3G) systems. The following are 2.5G systems:

- High-Speed Circuit-Switched Data (HSCSD): Even though most 2.5G systems implement packet switching, HSCSD continues support for circuit-switched data. It offers a data rate of 115 kbps and is designed to enhance GSM networks. The access technology used is time division multiple access (TDMA). It provides support for Web browsing and file transfers.
- General Packet Radio Service (GPRS): GPRS offers a data rate of 168 kbps. It enhances the performance and transmission speeds of GSM. GPRS provides always-on connectivity, which means users do not have to reconnect to the network for each transmission. Because there is a maximum of eight slots to transmit calls on one device, it allows more than one transmission at one time; for example, a voice call and an incoming text message can be handled simultaneously.
- Enhanced Data Rates for GSM Evolution (EDGE): EDGE works in conjunction with GPRS and TDMA over GSM networks. Its offered data rate is 384 kbps. EDGE supports data communications while voice communications are supported using the technology on existing networks.

The third-generation (3G) wireless systems are designed to support high bit rate telecommunications. 3G systems are designed to meet the requirements of multimedia applications and Internet services. The bit rate offered ranges from 144 kbps for full mobility applications, 384 kbps for limited mobility applications in macro- and microcellular environments, and 2 Mbps for low-mobility applications in micro- and picocellular environments. A very useful service provided by 3G systems is an emergency service with the ability to identify a user's location within 125 m 67% of time. Figure 1.1 shows the evolution of wireless standards.

Initially, the International Telecommunication Union (ITU) intended to design a single 3G standard; however, due to a number of difficulties, it has ratified two 3G standards. The two standards are CDMA2000, which provides a bit rate of up to 2.4 Mbps, and wideband CDMA (WCDMA), which provides a bit rate of up to

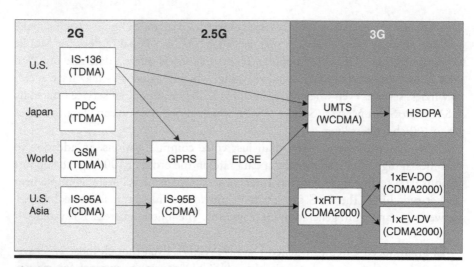

Figure 1.1 Evolution of wireless standards.

8 Mbps. The high bit rate enables new wireless services that can be classified into three categories:

1. Information retrieval: It permits location-aware applications to remotely download information from a corporate database.
2. Mobile commerce: It allows users to book a flight or pay bills.
3. General communication: It permits users to make or receive phone calls, send or receive messages, or activate an air conditioner at home.

Compound wireless service enables users to combine different types of services to carry out specialized functions. For example, you can take a photo using a camera phone and send it to a friend using the multimedia message service (MMS). A more useful application would be to combine a home alarm system with a wireless service so that when an intruder is detected, a photo of the intruder is captured by the surveillance camera and sent to the authorities, while the owner is alerted via mobile phone.

A compound service comprises a fundamental wireless service (one that cannot be partitioned into smaller identifiable services), a utility service (one that carries out a function within a particular compound service sequence), and possibly another compound service. For example, consider a courier service driver who has to deliver a document before a certain deadline and he has to find the fastest and least congested route to his destination. He makes use of a route planning application on the wireless terminal in his van, which consists of three fundamental wireless services:

1. A location service to determine the current location of the driver.
2. A travel route computation to determine the least congested and fastest route to his destination.

3. Traffic information retrieval to obtain traffic information from various sources.

The compound service consists of continuous iterations of these services: determine the current location and provide it to the wireless terminal, compute the least congested route from the current location to the destination, and retrieve the most updated traffic information. It involves executing step 1 and deciding whether to repeat step 2. Going back to step 1 is the utility service.

1.3 Summary

Mobile computing is an active area of research. Most applications available to users today are targeted at teenagers and yuppies and are mostly infotainment applications, for example, music downloads, friend locators, news updates. It will probably be a few more years before mobile enterprise applications appear on the market as there are many issues that need to be addressed to efficiently and effectively provide such applications. For this reason, several examples discussed in this book are based on ongoing and experimental work at various universities and research institutes. Therefore, a list of references, bibliographies, and online resources are provided at the end of each chapter so that readers may further explore the topic.

References

Liszka, K. J., M. A. Mackin, M. J. Lichter, D. W. York, D. Pillai, and D. S. Rosenbaum. 2004. Keeping a beat on the heart. *IEEE Pervasive Computing* 39(4):42.

Murray, S. 2004. Virtual fences: Herding cattle from home? *IEEE Pervasive Computing* 3(3):7.

Petrelli, D., E. Not, M. Zancanaro, C. Strapparava, and O. Stock. 2001. Modelling and adapting to context. *Personal and Ubiquitous Computing* 5:20.

Szewczyk, R., E. Osterweil, J. Polastre, M. Hamilton, A. Mainwaring, and D. Estrin. 2004. Habitat monitoring with sensor networks. *Communications of the ACM* 47(6):34.

Varshney, U., and R. Vetter. 2002. Mobile commerce: Framework, applications and networking support. *Mobile Networks and Applications* 7(2):185.

Weiser, M. 1991. The computer of the 21st century. *Scientific American*, September, 67.

Bibliography

Garlan, D., D. P. Siewiorek, A. Smailagic, and P. Steenkiste. 2002. Project Aura: Toward distraction-free pervasive Computing. *Pervasive Computing* 1(2):22.

Kobylarz, T. J. A. 2004. Beyond 3G: Compound wireless services. *Computer* 37(9):23.

Lorincz, K., D. J. Malan, T. R. Fulford-Jones, A. Nawoj, A. Clavel, V. Shnayder, G. Mainland, M. Welsh, and S. Moulton. 2004. Sensor networks for emergency response: Challenges and opportunities. *IEEE Pervasive Computing* 3(4):16.

Voth, D. 2004. Wearable Aid for the Visually Impaired. *IEEE Pervasive Computing* 3(3):6.
Weiser, M. 1993. Some computer science issues in ubiquitous computing. *Communications of the ACM* 36(7):75.

Online Resources

Aura Project. http://www-2.cs.cmu.edu/~aura/ (Accessed February 5, 2007).
CodeBlue Project. http://www.eecs.harvard.edu/~mdw/proj/vitaldust/ (Accessed February 5, 2007).
Wearable Computing at MIT. http://www.media.mit.edu/wearables/ (Accessed February 5, 2007).

Chapter 2

Cellular Network Architecture

When you subscribe to a mobile telephony service, your information is stored in a database called a home location register (HLR). The HLR plays an important role in providing you with the services offered by your service provider. In this chapter, you will learn about the cellular network architecture and the protocols involved in providing various services.

In a cellular network, a geographical region is divided into service areas called "cells." A cell is represented as a hexagon (Figure 2.1). At the center of a cell, is a base transceiver station (BTS) that serves users within the cell. A cluster of BTSs forms what is termed as Node B. Each cell is allocated a certain number of channels operating at a certain frequency. Channels used for transmission from the BTS to a mobile station (MS) are termed forward channels, and channels for transmission from a MS to a BTS are termed reverse channels. A few of these channels are reserved to send control data, for example, registration requests, call requests, authentications, paging to find a mobile user. A reverse control channel (RCC) and a forward control channel (FCC) are examples of channels used for transmission of control data.

A cellular network consists of a hierarchy of the following entities (Figure 2.2):

■ MS: It is a device used to communicate in the cellular network, for example, a mobile phone.
■ Base transceiver station: It is a BS consisting of a transceiver that receives or transmits signals over the radio interface. It serves one cell only.

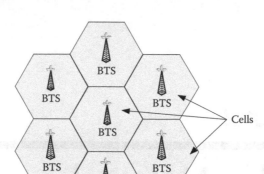

Figure 2.1   Cellular network architecture.

- ■ BS controller (BSC): It controls one or more BTSs and is under one mobile switching center (MSC).
- ■ Mobile switching center: It sets up and maintains calls made in the network. A MSC connects the cellular network to the fixed telephone network infrastructure (i.e., the public switched telephone network [PSTN]). It performs all switching and signaling functions for MSs located in its area.
- ■ BS subsystem (BSS): It consists of one BSC and one or more BTSs. The radio equipment of a BSS may support one or more cells.

Each entity performs a specific function. To examine how each entity performs its function, consider what happens when you dial your friend's home number. When you press on the call button, your mobile phone sends a call request to your BTS using a special channel (i.e., the RCC). The BTS forwards the request to the MSC. The MSC validates the request to make sure that you are authorized to use

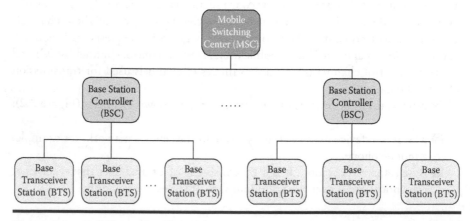

Figure 2.2   Cellular network entities.

the service and uses your friend's number to make a connection via the PSTN. A switch in the PSTN sets up a connection between your MSC and your friend.

A telephone number on a PSTN contains location information that is used by a switch to establish a connection between two subscribers; for example, 03 7967 6300 is the telephone number in Kuala Lumpur. On the contrary, a mobile user does not stay at a fixed location; hence, the mobile phone number cannot be used to determine a mobile user's current location. To deliver a call to a mobile user, the cellular network has to determine the user's location. The process of locating a user is termed paging.

Let us say that your friend dials your mobile phone number. When the request for connection establishment arrives at the MSC, it sends a broadcast message to all BSs under its control. The BSs then broadcast a paging message, which contains your mobile phone number, on all FCCs. Your mobile phone scans the FCC periodically to check if there is a paging message for you. When it detects the paging message, it acknowledges its presence in the cell by sending a message on the RCC. When the MSC receives the acknowledgment via the BS, it instructs the BS to allocate an unused voice channel for you. A data message is sent on the FCC to your phone to instruct it to ring. When you accept the call, a connection is established between you and your friend. When you have finished your conversation and hang up, the channel that was allocated to you is freed and can now be allocated to another user.

As you move from one location to another, the network needs to keep track of your location so that it can deliver calls to you. This is achieved using location registration and updates. The following sections discuss how this is achieved. We start by discussing the network architecture, followed by a discussion on the procedures and protocols required to support the services you use.

2.1   UMTS Architecture

The Universal Mobile Telecommunications Service ([UMTS] 3GPPTS23.101 V5.0.1 2004) is a 3G broadband, packet-based transmission of text, digitized voice, video, and multimedia services to mobile computer and phone users, regardless of their location in the world. It offers a data rate of up to 2 Mbps. Once UMTS is fully implemented and available, it will allow users to be constantly connected to the Internet as they roam, and users will have access to the same set of services and the same capabilities regardless of where they are.

UMTS is divided into two domains (Figure 2.3): the user equipment domain and the infrastructure domain. The interface between the two domains is the Uu reference point. The user equipment domain consists of various types of equipment with varying levels of functionality. The user uses a user equipment (UE) device to access UMTS services, such as a PDA or pocket PC. It has a radio interface to access the network. The user equipment domain is divided into two subdomains:

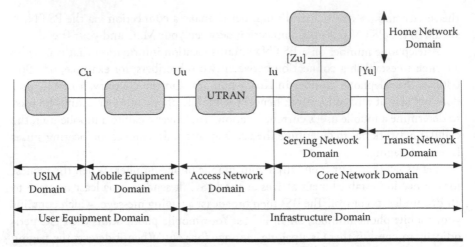

Figure 2.3   UMTS domains and reference points. (3GPP TS 23.101 V5.0.1, January 2004. Used with permission.)

1. User service identity module (USIM) domain: It contains data and procedures that unambiguously and securely identify itself. Because a device is associated with a specific user, it allows the ID of the user. The reference point between the USIM and mobile equipment (ME) domains is termed Cu.
2. ME domain: It contains applications and performs radio transmission.

The infrastructure domain is subdivided into two subdomains:

1. Access network domain: It consists of physical entities that manage the resources of the access network. It provides users with a mechanism to access the core network domain. The interface between this domain and the core network domain is termed Iu.
2. Core network domain: It consists of physical entities that provide support for the network features and telecommunication services (e.g., management of user location information, switching mechanism for signaling). It is further divided into three subsubdomains:
   a. Serving network domain: It is where the user access to the access network domain is connected. It represents the core network functions that are local to the user's access point (AP) and the location changes when the user moves. It is responsible for routing calls and transporting user data or information from source to destination. It interacts with the home domain to cater for user-specific data or services. The interface between this domain and the home network domain is termed Zu. It interacts with the transit domain for non-user-specific data or services. The interface between this domain and the transit network domain is termed Yu.

b. Home network domain: It represents the core network functions that are conducted at a permanent location regardless of the location of the user's AP. The USIM is related to this domain. It contains user-specific data and is responsible for the management of subscription information. It may also handle home-specific services not offered by the serving network domain.

c. Transit network domain: It is located on the communication path between the serving network domain and the remote party. If the remote party is located in the same network as the originating UE, no instance of the transit domain is activated.

2.1.1 UMTS Strata

Four UMTS strata are defined: transport stratum, serving stratum, home stratum, and application stratum. The home stratum involves domains shown in Figure 2.4, and the application stratum involves the domains shown in Figure 2.5. The serving and transport strata involve domains shown in both Figure 2.4 and Figure 2.5.

1. Transport stratum: Supports the transport of user data and network control signaling from other strata. It also provides the mechanism for error correction and recovery, data encryption, adaptation of data to use the supported physical interface, and transcoding of data to more efficiently use the radio interface. The transport stratum includes the access stratum, which consists of parts of both the infrastructure and UE. The protocols between these parts are specific to the access technique. It provides services related to data transmission over the radio interface and the management of the radio interface to other parts of UMTS. It includes two protocols: mobile termination–access network (MT-AN) and access network–serving network (AN-SN). The MT-AN protocol supports the transfer of radio-related information to coordinate the use of radio resources between the MT and access network. The AN-SN protocol supports the access from the serving network to the resources provided by the access network.

2. Serving stratum: Consists of protocols and functions to route and transmit user or network data from source to destination, which may be in the same or different networks. Telecommunication services functions are located in this stratum. It consists of three protocols: USIM–mobile termination (USIM-MT), mobile termination–serving network (MT-SN), and terminal equipment–mobile termination (TE-MT). The USIM-MT protocol supports access to subscriber-specific information to support functions in the UE domain. The MT-SN protocol supports access from the mobile terminal (MT) to the services provided by the serving network domain. The TE-MT protocol supports exchange of control information between the TE and MT.

3. Home stratum: Composed of protocols and functions to handle the storage of subscription data and home network–specific services. It also consists

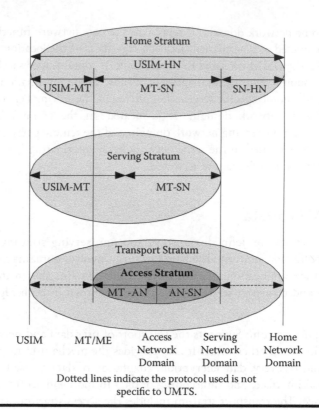

Figure 2.4   UMTS strata and the functional flows among USIM, MT/ME, access network, serving networks, and home networks domains. (3GPP TS 23.101 V5.0.1, January 2004. Used with permission.)

of functions to allow other domains to act on behalf of the home network. Among the functions provided are subscription data management, billing and charging, mobility management, and authentication. It consists of four protocols: USIM–home network (USIM-HN), USIM-MT, MT-SN, and serving network–home network (SN-HN). The USIM-HN protocol supports the coordination of subscriber-specific information between USIM and the home network. The USIM-MT protocol provides the MT access to user-specific data and resources required to perform action on behalf of the home network. The MT-SN protocol supports user-specific data exchanges between MT and the serving network. The SN-HN protocol provides the serving network with access to home network data and resources required to perform its actions on behalf of the home network, such as supporting user communications.

4. Application stratum: Represents the application process provided to end users. It provides the end-to-end protocols and functions to make use of services provided by the home, serving and transport strata, and the infrastructure to sup-

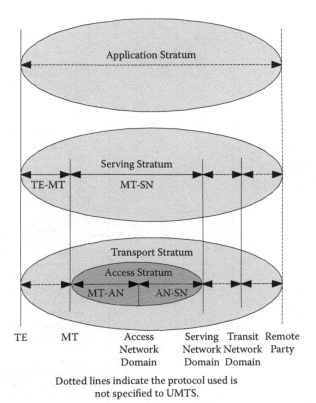

Figure 2.5   UMTS strata and the functional flows among TE, MT, access network, serving network, transit network domains, and the remote party. (3GPP TS 23.101 V5.0.1, January 2004. Used with permission.)

port services or value-added services. The protocols and functions may be the ones defined by GSM/UMTS standards or may be outside the UMTS standard. End-to-end functions are applications consumed by users at the edge of or outside the overall network and may be accessed by authorized users.

2.1.2   The Physical Layer

The physical layer offers services to the upper layers by defining the transport channel (3GPP TS25.211 V6.1.0 2004). A transport channel defines how data is transported over the air. It is divided into two groups: dedicated transport and common transport channels (Figure 2.6). There is only one type of dedicated transport channel, namely the dedicated channel (DCH). DCH is a downlink or uplink transport channel that is transmitted over the entire cell or over a part of the cell.

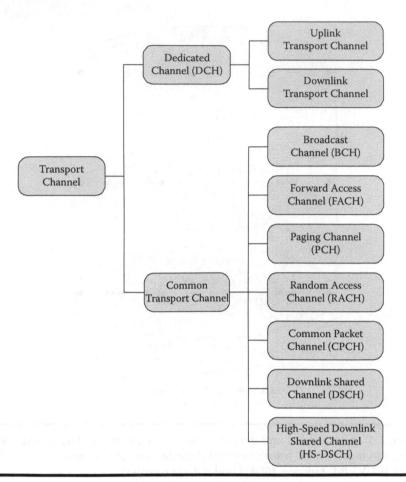

Figure 2.6   Classification of transport channel.

There are seven types of common transport channels:

1. Broadcast channel (BCH): A downlink channel to broadcast system- and cell-specific information. It is transmitted over the entire cell.
2. Forward access channel (FACH): A downlink channel that is transmitted over the entire cell.
3. Paging channel (PCH): A downlink channel that is transmitted over the entire cell. It is associated with the transmission of paging indicators to support efficient sleep mode procedures.
4. Random access channel (RACH): An uplink channel that is received from the entire cell. It is characterized by a collision risk and is transmitted using open loop power control.
5. Common packet channel (CPCH): An uplink channel associated with a downlink-dedicated channel that provides power control and CPCH control

command for the uplink CPCH. It is characterized by initial collision risk and is transmitted using inner loop power control.

6. Downlink shared channel (DSCH): A downlink channel shared by several UEs. It is associated with one or more downlink DCHs. It is transmitted over the entire cell or part of the cell.

7. High-speed downlink shared channel (HS-DSCH): A downlink channel shared by several UEs. It is associated with one downlink dedicated physical channel (DPCH) or several high-speed shared control channels (HS-SCCHs). It is transmitted over the entire cell or part of a cell.

Transport channels are mapped to physical channels. A physical channel is defined by its carrier frequency, scrambling code, channelization code, duration, and the relative phase. Physical channels are grouped into uplink physical channels and downlink physical channels (Figure 2.7).

There are three types of dedicated uplink physical channels:

1. Uplink dedicated physical data channel (uplink DPDCH): Used to carry the DCH transport channel. There may be zero, one, or more uplink DPSCHs on each radio link.

2. Uplink dedicated physical control channel (uplink DPCCH): Used to carry control information generated at layer 1. The control information consists of pilot bits, transmission power-control commands, feedback information, and an optional transport-format combination indicator.

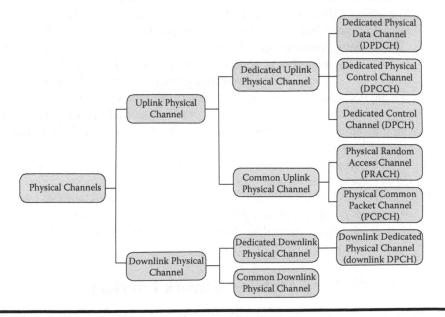

Figure 2.7   Classification of physical channel.

3. Uplink DPCH: Associated with HS-DSCH.

There are two types of common uplink physical channels:

1. Physical random access channel (PRACH): Used to carry RACH.
2. Physical common packet channel: Used to carry the CPCH.

The downlink physical channel is classified into dedicated downlink physical channel and the common downlink physical channel. There is only one type of dedicated downlink physical channel—the downlink dedicated physical channel (downlink DPCH). Downlink DPCH is a time multiplex of a downlink DPDCH and a downlink DPCCH.

There are 12 types of common downlink physical channels:

1. Common pilot channel (CPICH): A fixed rate 30-kbps channel that carries predefined bit sequences. There are two types of CPICH: primary CPICH and secondary CPICH. They differ in terms of usage and the limitations placed on their physical features.
2. Primary common control physical channel: A fixed rate channel of 30 kbps and is used to carry BCH.
3. Secondary common control physical channel (S-CCPCH): Used to carry FACH and paging channel (PCH).
4. Synchronization channel (SCH): A downlink signal used for cell search and consists of two subchannels: primary SCH and secondary SCH.
5. Physical downlink shared channel: Used to carry DSCH.
6. Acquisition indicator channel (AICH): A fixed rate physical channel used to carry acquisition indicators (AIs) that correspond to signatures on PRACH.
7. CPCH access preamble acquisition indicator channel: A fixed rate physical channel that carries the AP acquisition indicators (API) of CPCH.
8. CPCH collision detection/channel assignment indicator channel: A fixed rate physical channel that carries CD indicator if the CA is inactive.
9. Paging indicator channel: A fixed rate physical channel that carries the paging indicators. It is associated with S-CCPCH to which a PCH transport channel is mapped.
10. CPCH status indicator channel: A fixed rate physical channel that carries CPCH status information.
11. Shared control channel (HS-SCCH): A fixed rate downlink physical channel that carries downlink signaling related to HS-DSCH transmission.
12. High-speed physical downlink shared channel: Carries the HS-DSCH.

2.2 Public Land Mobile Network Interfaces

A public land mobile network (PLMN) interface performs functions as listed in Table 2.1. Figure 2.8 depicts the configuration and interfaces of PLMNs.

Table 2.1	A list of PLMN interfaces.	
Interface	*Between*	*Purpose*
A	MSC and BSS	It carries information concerning BSS management, call handling, and mobility management.
Abis	BSC and BTS	It supports the services offered by the network to the subscribers. Allows control of the radio equipment and RF allocation in the BTS.
B	MSC and associated VLR	It is used by the MSC when it needs to interrogate the VLR regarding a MS currently in its area and to inform its VLR of a location update procedure initiated by a MS. It is also used when the MSC needs to inform the HLR, via the VLR, to update data modified by a subscriber regarding a specific supplementary service.
C	HLR and MSC	To interrogate the HLR to obtain routing information for a subscriber.
D	HLR and VLR	To exchange data related to a MS's location and to the management of the subscriber. The VLR informs the HLR of the location of a MS managed by the latter and provides it with the MS's roaming number. The HLR sends the VLR the data required to support the service to the subscriber. Also, when the HLR needs to instruct a previous VLR, it sends the data required to cancel the location registration of a subscriber.
E	MSC and MSC	When a MS moves from one MSC area to another MSC area during a call, a handover procedure has to be initiated to maintain the connection. The MSCs exchange data to execute the procedure. After the procedure is completed, the MSCs exchange information to transfer A-interface signaling as necessary. It is also used to transfer a short message between the MSC serving the MS and the MSC that acts as the interface to the short message service center (SC).
F	MSC and EIR	It is used for data exchange so that the EIR can verify the status of IMEI retrieved for the MS.
G	VLR and VLR	It is used during registration procedure when a MS moves from one VLR area to another.
H	HLR and AuC	It is used by the HLR to request data from AuC to authenticate and cipher data for a subscriber.
I	MSC and associated group call register	It is used by the MSC to retrieve data related to a requested voice group call or broadcast call.

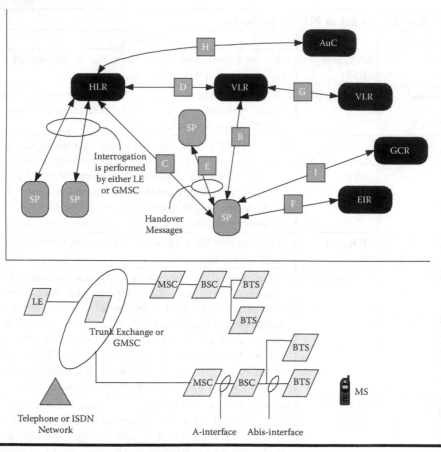

Figure 2.8 PLMN configuration and interfaces. (GSM 03.02 V5.3.0. January 1998. Used with permission.)

2.3 User Authentication

When you switch on your MS, it sends a registration request to the BS. Upon receiving the request, the BS executes the authentication procedure to ensure that the request is from a valid user. Once you are authenticated, you may access the services offered by your service provider.

An authentication center (AuC) is associated with a HLR and stores an identity key for each subscriber registered with the associated HLR. The key is used to generate the data used to authenticate the international MS identity (IMSI) and a key to cipher communication over the radio path between the MS and the network. An AuC only communicates with its associated HLR.

The equipment identity register (EIR) is a database that store the international MS equipment identity (IMEI) used in the system. ME may be classified as white listed, gray listed or black listed. At a minimum, an EIR contains a white list

Equipment that has been reported as stolen is classified as black listed and is not allowed to access the network.

2.4   Frequency Reuse

An important concept in cellular networks is frequency reuse. Because there are a limited number of available channels, frequency reuse makes it possible to support more users with limited resources. Each cell is allocated a set of channels. Adjacent cells are assigned a completely different set of channels to avoid cochannel interference. A footprint is the actual radio coverage of a cell and is determined from field measurements or propagation prediction models. Because the antenna of a BS is designed to cover only the cell it serves, the same set of channels can be assigned to two nonadjacent cells provided the distance between the two cells is large enough to keep interference levels within tolerable limits. The process of assigning the same set of channels to different cells is called frequency reuse (also termed frequency planning). In Figure 2.9, cells that are assigned the same set of channels are labeled with the same letter.

2.5   Channel Assignment

There are two types of channel assignment strategies: fixed channel assignment and dynamic channel assignment. In the fixed channel assignment strategy, a predetermined set of channels is allocated to a cell. A call request by a user is only served if

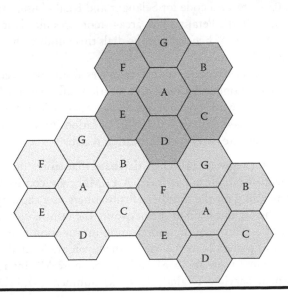

Figure 2.9   Frequency reuse.

there is an unused channel. Otherwise, the call is blocked. A more flexible variation of this strategy allows a cell to borrow channels from a neighboring cell if all of its channels have been assigned to users. The borrowing process is supervised by the MSC.

In a dynamic channel assignment strategy, channels are not allocated permanently to a cell. Each time there is a call request, the BS requests a channel from the MSC. The MSC allocates a channel after considering factors such as the probability of future call blocking, the frequency of use of the candidate cell, and the reuse distance of the channel. The MSC allocates a frequency provided it is not currently in use in the same cell or in any neighboring cells that would result in cochannel interference. The advantages of a dynamic strategy are that it reduces the probability of call blocking due to its flexibility, and that all available channels are accessible to all cells. A drawback of a dynamic strategy is that it is more complex because the MSC has to collect real-time data on channel occupancy, traffic distribution, and radio signal strength indication of all channels continuously. The amount of data collected and used in the channel allocation process increases storage and computational load on the system. The drawbacks are, however, compensated by increased channel utilization and reduced call blocking probability.

2.6 Location Registration and Update

If you are using a landline telephone, the location information is embedded in the telephone number. For example, location information is embedded in the number 03 7967 6300: 03 is the area code for Selangor and Kuala Lumpur; 7967 tells you that the subscriber is in the Petaling Jaya area—more specifically it is a number for the University of Malaya. Whenever anyone dials this number, the switches will set up a connection to the campus.

Conversely, because a mobile subscriber moves from one location to another, the mobile telephone number does not give location information. To deliver a call to a subscriber, the network operator needs to keep track of the location of all subscribers (ETSI TS 100 530 V7.0.0 1998). Location information is stored in a location register. There are two types of location registers:

1. Home location register (HLR): A database that stores information for the management of mobile subscribers. A PLMN may consist of one or more HLRs, depending on the number of subscribers, the capacity of the equipment, and the organization of the network. The information stored in HLRs are subscription information and location information to enable the charging and routing of calls to the MSC where you are located. Each subscription is associated with an IMSI and one or more MS international ISDN (International Services Digital Network) numbers (MSISDN). The HLR

may also store other information such as service restriction and supplementary services.

2. Visitor location register (VLR): A database that stores information required to handle call requests or deliveries made or received by subscribers roaming in its area. A VLR area is the part of the network controlled by a VLR and may consist of one or more MSC areas. A MSC area consists of all BSs under the control of the MSC and may consist of one or more location areas. The information stored in VLRs are the IMSI, the MSISDN, the MS roaming number, the temporary MS identity, the local MS identity and the location area where you are registered.

When is a location update triggered? It is triggered only when you move out of a location area and into another location area. A location area (LA) consists of one or more cells and is associated with a LA ID (LAI). When you roam into a new location area, your MS initiates a registration procedure. When the MSC in charge of the area notices the registration, it transfers the LAI of your location to the VLR. If your MS is not yet registered, the VLR and HLR exchange information to allow the proper handling of calls to you.

A service area is an area in which you can be reached by other mobile or fixed subscribers without them knowing your actual location. It may consist of several PLMNs.

When you power on your MS, it carries out an explicit IMSI attach operation to indicate to the PLMN that it has entered an active state. An explicit IMSI detach is carried out when you power down your MS (i.e., it enters an inactive state).

When a MS roams in a foreign network, the MSC passes information update messages between the MS and the VLR. An implicit detach timer is associated with the MS. This timer is derived from the periodic location updating timer. The VLR executes an implicit detach operation to mark a MS as detached when there has been no successful contact between the MS and the network for a period specified by the implicit detach timer. When a radio connection is established, the implicit detach timer is suspended and prevented from triggering an implicit detach. When the radio connection is released, the timer is reset and restarted.

2.7   Handover Procedures

Handover (also handoff) procedures (ETSI 100 527 V7.0.0 1998) ensure that the connection to a MS is maintained when it moves from one cell to another. Let us say you are in a train and you are talking to your friend using your mobile phone. As the train moves, it crosses the cell boundary. When this happens, the BS in your current cell has to handover your connection to the BS in the cell that you are moving into to maintain the connection between you and your friend; otherwise, you will be disconnected. This process is handled by the handover procedure. A part of the procedure involves the allocation of a new channel for you in the new cell and the release of the channel allocated to you in your current cell. Figure 2.10 shows neighboring cells to which the call may be handed over.

How does the network decide when it is time to handover the connection? Handover is initiated based on radio subsystem criteria (e.g., radio frequency [RF] level, signal quality) and network directed criteria (e.g., current traffic load in the cell). Let us say the handover decision is made based on RF level. The MS takes radio measurements from neighboring cells and reports it to the serving cell periodically. When the network determines that a handover is required, the handover procedure is initiated.

There are two types of handover:

1. Intra-MSC handover: Handover between BSs under the control of the same MSC, including inter- and intra-BSS handovers.
2. Inter-MSC handover: Handover between BSs under the control of different MSCs. This is further classified into three procedures:
   a. Basic inter-MSC handover procedure: A MS is handed over from a controlling MSC (MSC-A) to another MSC (MSC-B).
   b. Subsequent inter-MSC procedure: A MS is handed over from a MSC-B to a third MSC (MSC-B').
   c. Subsequent inter-MSC handback: A MS is handed back from MSC-B to MSC-A.

Table 2.2 lists the messages exchanged during a handover procedure.

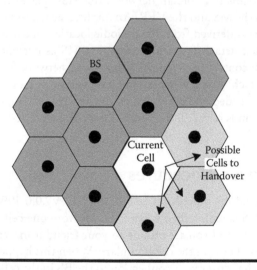

Figure 2.10   Initiation of handover procedure.

Table 2.2   A list of messages sent in the handover procedure.

Message	From → To	Purpose
A-Clear-Command	MSC-A → BSS-A MSC-B → BSS-A	To begin the release of resources allocated for the MS.
A-Clear-Complete	BSS-A → MSC-A BSS-A → MSC-B	To indicate that the clearing of resources allocated for the MS has been completed.
ACM (SS7)	MSC-B → MSC-A	To establish a circuit between MSC-A and MSC-B.
A-Handover-Command	MSC-A → BSS-A MSC-B → BSS-A	To initiate the process of instructing the MS to tune to the new channel.
A-Handover-Complete	BSS-B → MSC-A BSS-B → MSC-B	Sent when the MS has communicated successfully with BSS-B.
A-Handover-Detect	BSS-B → MSC-A BSS-B → MSC-B	To be sent to indicate that the MS has been captured.
A-Handover-Request	MSC-A → BBS-B MSC-B → BBS-B	To initiate the handover process. It contains information required by MSC-B to allocate a new channel.
A-Handover-Request-Ack	BSS-B → MSC-A BSS-B → MSC-B	To acknowledge that a channel has been allocated for the MS. It contains the radio resource definition that is later forwarded to the MS.
A-Handover-Required	BSS-A → MSC-A BSS-A → MSC-B	To inform MSC-A that a handover is required.
IAM (SS7)	MSC-A → MSC-B	To establish a circuit between MSC-A and MSC-B.
MAP-Allocate-Handover-Number request	MSC-B → VLR-B	To obtain a handover number used for routing the connection from MSC-A to MSC-B.
MAP-Prepare-Handover request	MSC-A → MSC-B	To identify the cell to which the MS is to be handed over.
MAP-Prepare-Handover response	MSC-B → MSC-A	Returned if a traffic channel is available. Contains the complete A-Handover-Request-Ack or A-Handover-Failure it received from BSS-B.

MAP-Prepare-Subsequent-Handover request	MSC-B → MSC-A	It contains the new MSC number (MSC-A), new cell identity, and the complete A-Handover-Required or A-Handover-Request message.
MAP-Prepare-Subsequent-Handover response	MSC-A → MSC-B	Returned after MSC-A has assigned a new radio channel. It contains the complete A-Handover-Request-Ack.
MAP-Process-Access-Signaling request	MSC-B → MSC-A	To forward A-Handover-Detect received by MSC-B from the correct MS. It is also used to send traffic channel allocation results.
MAP-Send-End-Signal request	MSC-B → MSC-A MSC-B′ → MSC-A	To forward A-Handover-Complete to MSC-A.
MAP-Send-End-Signal response	MSC-A → MSC-B	To terminate the dialogue between them.
MAP-Send-Handover-Report request	VLR-B → MSC-B	It contains the handover number requested by MSC-B.
RI-Handover-Access	MS → BSS-B	Contains the handover reference number so that BSS-B can ensure that everything is as expected and that the correct MS has been captured.
RI-Handover-Command	BSS-A → MS	To instruct the MS to tune in to the new channel based on the handover reference number provided by BSS-B.
RI-Handover-Complete	MS → BSS-B	Sent when MSS has communicated successfully with BSS-B.

2.7.1 Intra-MSC Handover

The MSC-A controls the call, mobility management of the MS, and radio resources before, during, and after the handover. The BS system application part (BSSAP) procedures are initiated and handled by MSC-A. MSC-A initiates and controls the handover from its initiation until its completion. There are two types of handover:

1. Internal handover: Takes place between channels in a cell or cells controlled by a single BSS without reference to the MSC, but the MSC may be informed of the handover.

2. External handover: Takes place between channels on the same cells or between cells on the same BSS that is controlled by the MSC.

When the network detects that a handover is required, the following procedure is initiated. The messages passed during this procedure are shown in Figure 2.11. When BSS-A (the current BSS) determines that the MS should be handed over, it sends A-Handover-Required to MSC-A. This message contains a list of cells to which the MS can be handed over, in order of preference, based on the criteria defined by the network operator. When MSC-A receives the message, it begins the process of handing over the MS to BSS-B (the new BSS). MSC-A generates A-Handover-Request to BSS-B. When BSS-B receives the message, it allocates its radio resource to the MS. Once the allocation is completed, it sends an acknowledgment, A-Handover-Request-Ack, to MSC-A. When MSC-A receives the message, it initiates the process of instructing the MS to tune to the new channel by sending A-Handover-Command to BSS-A, which in turn, sends a RI-Handover-Command to the MS. This message contains a handover reference number allocated by BSS-B to the MS. The MS uses this reference number to access the new radio resource. This reference number is also included in a RI-Handover-Access sent by the MS to BSS-B so that BSS-B can check to ensure that it is as expected and the correct MS has been captured.

After BSS-B determines that it is correct, it sends A-Handover-Detect to MSC-A. When the MS successfully communicates with BSS-B, it sends RI-Handover-Complete to BSS-B, which in turn, sends A-Handover-Complete to MSC-A. Upon receiving the message, MSC-A sends A-Clear-Command to BSS-A so that BSS-A may begin the release of the resource allocated for the MS by BSS-A.

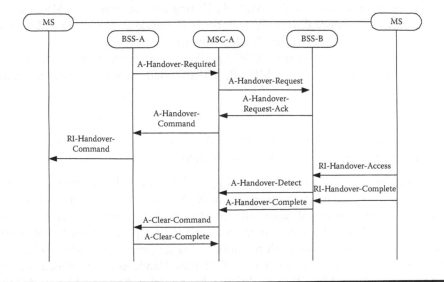

Figure 2.11   Basic external intra-MSC handover procedure.

If a failure occurs during the handover, MSC-A may take any of the following actions:

- Retry the handover to the same cell.
- Select the next cell from the list in A-Handover-Required and try to handover to another cell.
- Wait for the next A-Handover-Required message.
- Send A-Handover-Required-Reject to BSS-A if A-Handover-Command has not been sent already.

The action to be taken depends on whether the failure occurs before or after A-Handover-Command has been sent. In any case, the existing connection to the MS is not cleared. During the process of switching to a new radio resource, there may be a short period when the MS is not in communication with the network. Any messages for the MS are queued at MSC-A and are delivered to it once communication resumes.

2.7.2 Basic Inter-MSC Handover

MSC-A controls the call and the mobility management before, during, and after a basic or subsequent handover. The BSSAP procedures are initiated and handled by MSC-A, if required. During a subsequent inter-MSC handback, MSC-A acts as a BSS to MSC-B. MSC-B controls the handover procedure until the termination of radio resource allocation in MSC-A. After radio channels have been allocated, all handover messages terminate in MSC-A. During a subsequent inter-MSC procedure, MSC-A acts toward MSC-B′ as in the basic handover procedure and toward MSC-B as in the subsequent handback procedure.

There are two types of inter-MSC handover: one that requires a circuit connection between MSC-A and MSC-B, and one that does not require a circuit connection.

2.7.2.1 Basic Handover Requiring a Circuit Connection between MSC-A and MSC-B

The handover is initiated by BSS-A sending the A-Handover-Required message (Figure 2.12). Upon receiving the message, MSC-A sends a MAP-Prepare-Handover request (MAP: mobile application part) to MSC-B. This message identifies the cell to which the MS is to be handed over and contains a complete A-Handover-Request. A-Handover-Request contains all the information required by MSC-B to allocate a new radio channel. MSC-B sends a MAP-Allocate-Handover-Number request to its associated VLR (i.e., VLR-B) to obtain a handover number that is used for routing the connection of the call from MSC-A to MSC-B. Upon receiving the handover num-

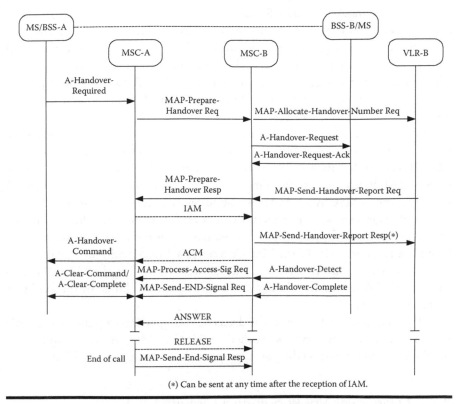

Figure 2.12 Basic handover procedure requiring circuit connection.

ber in the MAP-Send-Handover-Report request message, MSC-B returns a MAP-Prepare-Handover response to MSC-A.

If a traffic channel is available at MSC-B, it sends a MAP-Prepare-Handover response to MSC-A. This message contains the complete A-Handover-Request-Ack that it receives from BSS-B. A-Handover-Request-Ack contains a radio resource definition that is later forwarded by BSS-A to the MS. Other traffic channel allocation results (A-Handover-Request-Ack or A-Handover-Failure) are sent to MSC-A using a MAP-Process-Access-Signaling request. If MSC-B is unable to allocate a channel, the MAP-Prepare-Handover response containing A-Handover-Failure is sent to MSC-A. The same is done if MSC-B detects a fault on the identity of the cell where the call is to be handed over. It is then up to MSC-A to decide on the action to take upon receiving negative responses or if the operation fails due to the expiry of the MAP-Prepare-Handover timer.

If MSC-A receives an error indication, it terminates the handover attempt. It may retry the handover using the cell identity list (if available) or it may reject the handover attempt. In case of failure, the existing connection to the MS is not cleared.

Upon receiving A-Handover-Request-Ack, MSC-A establishes a circuit between itself and MSC-B by using the initial address message (IAM) and address complete message (ACM) of Signaling System No. 7 (SS7). If the circuit between MSC-A and MSC-B cannot be established, MSC-A aborts the handover by sending an appropriate error message, for example, ABORT. MSC-B waits for the capturing of the MS on the radio path when ACM is sent, and MSC-A initiates the handover execution when ACM is received.

When MSC-B receives an acknowledgment from the correct MS (A-Handover-Detect or A-Handover-Complete), it forwards the acknowledgment to MSC-A. If MSC-B receives A-Handover-Detect from the correct MS, it forwards it to MSC-A using a MAP-Process-Access-Signaling request. If A-Handover-Complete is received, it is forwarded to MSC-A in a MAP-Send-End-Signal request. The circuit in MSC-A is through-connected when it receives A-Handover-Detect or A-Handover-Complete from MSC-B. The old radio channel is released upon the receipt of A-Handover-Complete from MSC-B. When MSC-A receives a MAP-Send-End-Signal request that contains A-Handover-Complete, the resource allocated to the MS at BSS-A is cleared. MSC-A retains overall call control until the call is cleared by the fixed subscriber of the MS.

MSC-A aborts the handover if it detects clearing or interruption of the radio path before the call has been established on MSC-B. MSC-A may terminate the handover at any time by sending an appropriate MAP message to MSC-B.

2.7.2.2 Handover Not Requiring a Circuit Connection between MSC-A and MSC-B

The difference between this procedure and the one explained in the previous section is the establishment of circuits between network entities and the handover number allocation (Figure 2.13). MSC-A sends a MAP-Prepare-Handover request to MSC-B, specifying that no handover number is required. MSC-B sends A-Handover-Request to BSS-B and allocates radio resources for the handover. It returns a MAP-Prepare-Handover response containing A-Handover-Request-Acknowledge or A-Handover-Failure received from BSS-B to MSC-A. The handover procedure proceeds as explained in section 2.7.2.1, except that there is no circuit connection establishment between MSC-A and MSC-B.

2.7.3 Subsequent Handover Requiring Circuit Connection between MSC-A and MSC-B

If a subscriber moves into another MSC area during the same call, a subsequent handover is triggered to continue the connection (Figure 2.14). There are two possibilities: the MS moves back into the area of MSC-A or it moves into the area of

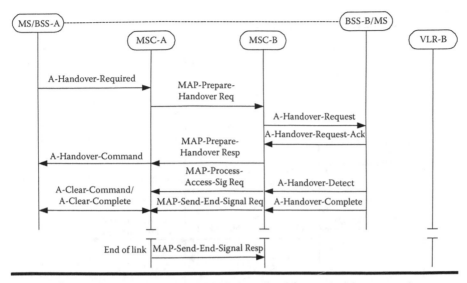

Figure 2.13 Basic handover procedure not requiring a circuit connection.

a third MSC (i.e. MSC-B'). The circuit between MSC-A and MSC-B is released after a successful handover.

2.7.3.1 Subsequent Handover from MSC-B to MSC-A

When MSC-B receives A-Handover-Required from BSS-A, it sends a MAP-Prepare-Sub-Handover request containing the new MSC number (MSC-A) and the cell identity where the call is to be handed over to MSC-A. The complete A-Handover-Required is included in this message. Upon receiving the message, MSC-A initiates a search for a free radio channel. When it has assigned a radio channel, it returns a MAP-Prepare-Sub-Handover response containing a complete A-Handover-Request-Ack to MSC-B.

2.7.3.2 Subsequent Handover from MSC-B to MSC-B'

This procedure is composed of two parts (Figure 2.15):

■ A subsequent handover from MSC-B to MSC-A
■ A basic handover from MSC-A to MSC-B'

MSC-B sends a Map-Prepare-Subsequent-Handover request to MSC-A that contains the new MSC number, target cell identity, and a complete A-Handover-Request. MSC-A triggers the basic handover procedure toward MSC-B'. Upon receiving an ACM message, MSC-A sends a MAP-Prepare-Subsequent-Handover response,

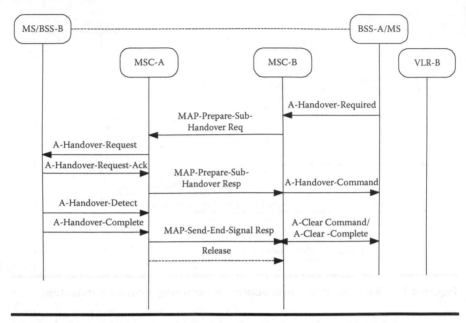

Figure 2.14   Subsequent handover from MSC-B to MSC-A with circuit connection.

containing the complete A-Handover-Request-Ack received from BSS-B′, to inform MSC-B that MSC-B′ has successfully allocated radio resources on BSS-B′. When MSC-B receives this message, it initiates the procedure on the radio path.

The handover procedure at MSC-A is completed when MSC-A receives a MAP-Send-End-Signal request from MSC-B′ containing A-Handover-Complete from BSS-B′, and the circuit between MSC-A and MSC-B is released. MSC-A sends MAP-Send-End-Signal response to MSC-B to terminate the dialogue between them. Subsequently, MSC-B releases the radio resources.

If any of the following occurs, MSC-A informs MSC-B of the failure by sending A-Handover-Failure that is included in the MAP-Prepare-Subsequent-Handover response:

- ■ MSC-B′ cannot allocate a radio channel.
- ■ A circuit cannot be established between MSC-A AND MSC-B′.
- ■ A fault is detected on the target cell identity.
- ■ The target cell identity in A-Handover-Request is not consistent with the target MSC number.

MSC-B has to maintain the existing connection with the MS. When the handover is completed, MSC-B′ becomes MSC-B.

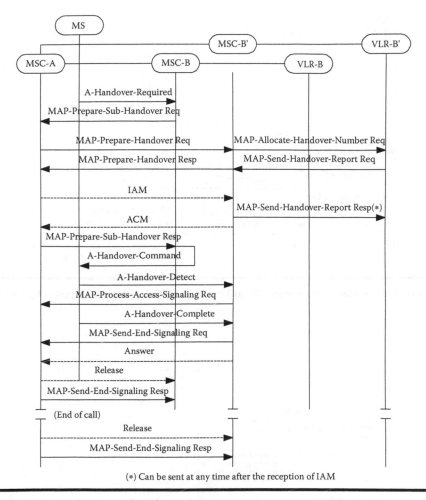

Figure 2.15 Subsequent handover from MSC-B to MSC-B' with circuit connection.

2.7.4 Subsequent Handover Not Requiring Circuit Connection between MSC-A and MSC-B

As in section 2.7.3, there are two possibilities: the MS moves back into the area of MSC-A, or it moves into the area of a third MSC (i.e., MSC-B').

Figure 2.16 shows the subsequent handover from MSC-B to MSC-A. The only difference between this procedure and the one shown in Figure 2.14 is that there is no circuit release between MSC-A and MSC-B'. Figure 2.17 shows the subsequent handover procedure from MSC-B to MSC-B'. This is different from the one shown in Figure 2.15 as there is no circuit and no handover number allocation signaling.

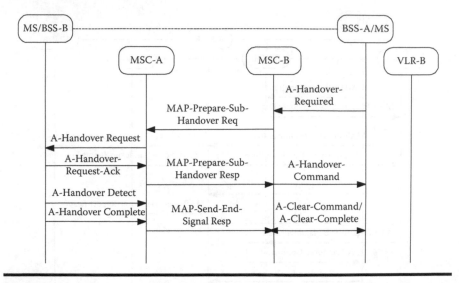

Figure 2.16   Handover procedure from MSC-B to MSC-A without a circuit connection.

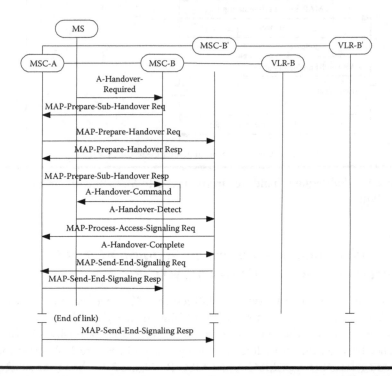

Figure 2.17   Subsequent handover procedure from MSC-B to MSC-B′ without circuit connection.

2.8 CDMA

CDMA is a spread spectrum technology used in 2G and 3G communications systems. It is used in ultrahigh frequency cellular telephone systems in the 800-MHz and 1.9-GHz bands. It is a multiplexing technique that allows several signals to share a transmission channel, thus, optimizing the use of available bandwidth. Transmissions from different stations occupy the entire frequency band at the same time. Different codes are used to produce the transmitted signals and to distinguish them. The code used by one station is orthogonal to the codes used by other stations.

An orthogonal code is similar to an object being orthogonal in a three-dimensional (3D) space, as illustrated in the following simplified examples. Two vectors are orthogonal if their inner product is 0. Let us say we have two vectors: (1, 2, 3) and (2, 2, -2). The product of the two vectors is $(1, 2, 3) \times (2, 2, -2) = 2 + 4 + (-6) = 0$. Two vectors are considered almost orthogonal if their product is close to 0: for example, $(1, 2, 3) \times (-3, 2, 0) = -3 + 4 + 0 = 1$. On the contrary, (2, 4, 6) is not orthogonal to (1, 3, 5) because their product is $(2, 4, 6) \times (1, 3, 5) = 2 + 12 + 30 = 44$.

A spread spectrum technique spreads the bandwidth needed to transmit data. Assume that user data is generated at R bps. Each station is assigned a unique binary pseudorandom sequence, also known as the chip values. Each bit of user data is transformed into G bits by multiplying the user bit value by G chip values. The result is a sequence that appears to be random. The spreading factor G is selected so that the transmitted signal occupies the entire frequency band of the medium. A chip is one bit of a digital sequence spread spectrum (DSSS) code. A chip rate of a code is the number of bits per second (chips per second) at which the code is transmitted or received.

The sequence G chip values are $(c_1, c_2, c_3, \ldots, c_G)$, where c_i takes a value of either +1 or -1. The -1 symbol represents bit 0, and +1 represents bit 1. When a station wants to transmit -1, it multiplies the -1 value by the sequence of G chip values. The resulting value is the signal transmitted to the receiver. When the signal arrives at the receiver, it is multiplied by G again and the resulting products are added. If the signal is received correctly, the resulting output correlator is either $+G$ or $-G$.

Example

Station A and station B use spreading sequences that are orthogonal to each other. Assume that $G = 4$ and the spreading sequence of stations A and B are (+1, -1, -1, +1) and (+1, -1, +1, -1), respectively. Let us say station A wants to transmit -1 (bit 0). Station A multiplies the symbol it wants to transmit by its spreading sequence

$$-1 \times (+1, -1, -1, +1) = (-1, +1, +1, -1).$$

The resulting symbol is transmitted to the receiver. When the receiver receives the signal, it calculates the product of the signal it receives with station A's spreading sequence

$$(-1, +1, +1, -1) \times (+1, -1, -1, +1) = (-1, -1, -1, -1) = -4 = -G.$$

The result of the calculation (i.e., $-G$) tells the receiver that it has decoded the signal correctly and that the received symbol is -1 (bit 0). If station A transmits +1, the result of the calculation should yield $+4 = +G$, which tells the receiver that the transmitted value is +1 (bit 1).

If station B receives the signal and tries the decode it using its spreading sequence, the result is

$$(-1, +1, +1, -1) \times (+1, -1, +1, -1) = -1 -1 +1 +1 = 0.$$

Station B is unable to determine whether the received symbol is +1 or -1. Only a transmitter and receivers with knowledge of the correct spreading sequence would be able to decode a received signal correctly.

CDMA is compatible with other cellular technologies, thus, allowing for nationwide roaming. The original CDMA standard is CDMA One and is still used in cellular telephone networks in the United States. The transmission speed is between 14.4 and 115 kbps.

There are a number of variations of the basic CDMA, such as WCDMA, time division CDMA, and CDMA2000. The ITU ratified two 3G standards, namely WCDMA and CDMA2000. WCDMA supports voice, images, data, and video. Utilizing the 5-MHz channel bandwidth, it is capable of carrying over 100 simultaneous voice calls and offers a data rate up to 8 Mbps.

CDMA2000 1xRTT (radio transmission technology standard, also referred to as IS-2000), CDMA2000 EV-DO (evolution–data optimized), and CDMA2000 EV-DV (evolution–data and voice) are CDMA2000 approved radio interfaces for ITU's IMT-2000 standard. CDMA operates in the 1.25 frequency band. CDMA2000 EV-DO offers a data rate of $4.9 \times N$–Mbps downlink, and $1.8 \times N$–Mbps uplink, where N is the number of 1.25-MHz carriers dedicated to the system. CDMA2000 EV-DV offers data rates of 3-Mbps downlink and 1.8-Mbps uplink.

2.9 The Move Toward 3G Networks

As 3G services are rolled out, there are a number of issues that need to be addressed. This section gives a glimpse of a few issues that may arise.

Subscribers expect the same services available to them in their home network to be available when they are roaming, and at the same level of service quality. To illustrate how this might not always be possible, consider this scenario. Suppose that a subscriber from Kuala Lumpur roams in Paris and she requests a list of restaurants in a 5-km vicinity. Assuming that the visited network is able to provide her with the information, in what language should the information be presented? Would it be presented in Malay (her preferred language), English (as she is a visitor in a foreign network), or in French (which she might not understand)? Even though it is possible to require the subscriber to reconfigure her preferred settings when roaming, doing so means that a seamless service is not being provided. Moreover, it might not always be possible to translate information into the user's preferred language.

In addition to the technical issues in supporting seamless roaming, there are also nontechnical issues; for example, from a legal perspective. The law regarding the encryption of data varies from country to country. What happens when a user crosses the border of a country that allows the use of strong encryption into one that imposes restrictions on its use? Copyright issues also arise regarding the distribution of information that might be restricted to certain geographical areas. For example, if an operator buys the right to broadcast a Formula One race in Asia Pacific, can they make it available to roaming customers in Europe? These issues must be kept in mind when designing new services.

Because a 3G network is implemented in stages, it is possible that a subscriber may roam from an area with 3G coverage to an area without 3G coverage. In this case, it should be possible to fall back onto the existing 2G network infrastructure, albeit at reduced functionality or quality of service (QoS). The subscriber should be moved back to the 3G network when coverage is available, but this should only be done when the coverage is stable to avoid oscillation between the two networks. This fallback approach gives rise to the issue of how billing should be handled. When a subscriber moves into a 2G network, should billing be adjusted to reflect the reduced level of service provided to the subscriber during this period? Doing so would increase the complexity of pricing and billing. How this is to be handled ought to be specified in the service level agreement between network operators.

The examples above hint at the complexity of the issues that have to be addressed. Although technical issues can be resolved given time and technological advances, nontechnical issues may require agreements between service providers in different countries or even between governments.

References

3GPP TS 23.101 V5.0.1. January 2004. Technical specification group services and system aspects general UMTS architecture. Technical Specification.

3GPP TS 25.211 V6.1.0. August 2004. Physical channels and mapping of transport channels onto physical channels (FDD). Technical Specification.

ETSI TS 100 527 V7.0.0. August 1998. Handover procedures. Technical Specification.

ETSI TS 100 530 V7.0.0. August 1998. Location registration procedures. Technical Specification.

GSM 03.02 V5.3.0. January 1998. Digital cellular telecommunications system (phase 2+): Network architecture. Technical Specification.

Jaseemuddin, M. 2003. An architecture for integrating UMTS and 802.11 WLAN networks. *Proc. of the 8th IEEE International Symposium on Computers and Communications (ISCC '03)* 2:716.

Bibliography

Jou, Y. 2000. Developments in third generation (3G) CDMA technology. *Proc. of the IEEE 6th International Symposium on Spread Spectrum Techniques and Application* 2:60–64.

Roos A., M. Hartman, and S. Dutnall. 2003. Critical issues for roaming in 3G. *IEEE Wireless Communications* 10(1):29.

Online Resources

Third Generation Partnership Project (3GPP). http://www.3gpp.org/ (Accessed February 5, 2007).

CDMA Hub. http://www.geocities.com/rahulscdmapage/ (Accessed March 12, 2007).

European Telecommunication Standard Institute (ETSI). http://www.etsi.org/ (Accessed February 5, 2007).

International Telecommunication Union (ITU). http://www.itu.int/home/imt.html (Accessed February 5, 2007).

Spread Spectrum Technology Tutorial. http://www.commsdesign.com/design_corner/ OEG20030506S0029 (Accessed March 12, 2007).

UMTS Forum. http://www.umts-forum.org/ (Accessed March 14, 2007).

UMTS World. http://www.umtsworld.com/ (Accessed March 14, 2007).

Chapter 3

Wireless Local Area Networks

A wireless local area network (WLAN) is attractive for several reasons. A WLAN is flexible and can be installed anywhere. For a small company, the expense of installing a fixed network infrastructure might not be justified. Regulations on historical buildings might not permit the laying of cables that is required for installing a wired network infrastructure. In an enterprise, it extends access to the corporate local area network (LAN).

Another popular use of WLANs is to provide broadband Internet access at public places such as cafés and airports. Areas covered by this service are known as hot spots. At conferences where there are high numbers of people convening for short periods, WLANs are used to provide participants with access to the Internet since: deploying a wired LAN for such short periods is not practical.

This chapter covers WLAN standards and protocols. Institute of Electrical and Electronic Engineers (IEEE) has played an active role in defining standards for WLANs, known as the 802.11 family.

3.1   IEEE 802.11 Standard

Work on the 802.11 standard commenced in 1989, and the final draft was released in 1997. 802.11 defines a basic service set (BSS) as comprising two or more fixed, portable, or moving nodes or stations that can communicate with each other in a limited geographical area. It defines two configurations: an infrastructure mode (Figure 3.1) with at least one central AP connected to the wired network, and an ad

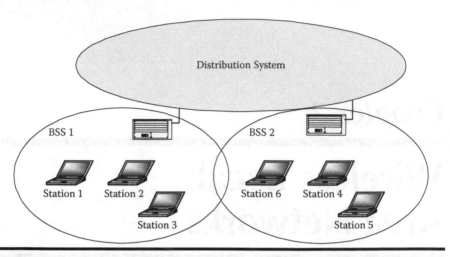

Figure 3.1   An infrastructure configuration.

hoc (Figure 3.2) or peer-to-peer mode where a set of wireless stations communicate directly with one another without requiring a central AP or a wired connection. In an infrastructure configuration, the BSS is connected to the distribution system, which is usually the wired LAN. A handover is triggered when a station moves from the coverage area of one AP to another. In an ad hoc configuration, all stations are independent and equivalent. Stations may broadcast packets in the wireless coverage area without accessing the Internet. This configuration is very useful when a quick network setup is required without the hassle of dealing with cables, such as at a conference venue.

802.11 defines the physical layer and the medium access (MAC) layer. It defines three RF physical layers that operate at the 2.4-GHz industrial, scientific, and medical (ISM) band: frequency hopping spread spectrum (FHSS), digital sequence spread spectrum (DSSS), and IR. FHSS and DSSS support a maximum bit rate of 2

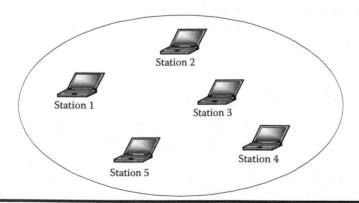

Figure 3.2   An ad hoc configuration.

Mbps in a clean environment and an optional 1 Mbps in a noisy environment. The IR physical layer offers 1 and 2 Mbps for receiving and 1 Mbps for transmitting. A 2 Mbps transmission rate is offered as an option. It employs a technique called diffuse IR transmission that does not require directed transmission. The IR physical layer is not as popular as the FHSS and DSSS because the 10 m range imposes a serious constraint on the system.

In a WLAN, the carrier sense multiple access with collision detection (CSMA/CD) is unsuitable due to the hidden station problem. To demonstrate why CSMA/CD is unsuitable, refer to Figure 3.3. Stations A and C are both within the transmission range of station B, but are not within range of each other. Stations A and C are hidden from each other. Station C is currently transmitting to station B. At the same time, station A would like to transmit to station B. If A is using CSMA/CD, it first has to determine if anyone else is transmitting to B by sensing the channel. Because C is not within the range of A, it is unable to detect C's transmission to B. It concludes that no one else is transmitting to B and starts its transmission, which collides with C's transmission at B.

To overcome the hidden station problem, the standard specifies a distributed coordination function (DCF) known as carrier sense multiple access with collision avoidance (CSMA/CA) as an access protocol at the MAC layer. When a station has data to send, it executes the following six steps:

1. It senses the channels.
2. If an idle channel is detected, the station continues sensing the channel for an additional duration, termed the DCF interframe space (DIFS), to avoid a collision in case two stations sense an idle channel at the same time. If the channel is still idle after a period of DIFS, the station may start transmission.
3. If the channel is busy, it defers transmission and enters a backoff state. The period following DIFS is called a contention window (CW) and consists of a predetermined number of slots.
4. Upon entering a backoff state, the station randomly selects a slot in the CW and continuously senses the channel for a period up to the selected slot.
5. If it does not detect any transmission, it captures the medium by starting to transmit its data.
6. If during this period it detects a transmission from another station, it enters a backoff state again and the whole process is repeated.

Station A   Station B   Station C

Figure 3.3   The hidden station problem.

Figure 3.4 demonstrates this operation. The receiving station sends an acknowledgment (ACK) for each frame received successfully. High-priority frames (e.g., ACK, request to send [RTS], clear to send [CTS]) need only wait for a period of short interframe space (SIFS) before contending for the channel, where SIFS is less than DIFS.

To reduce the hidden station problem, a RTS-CTS mechanism is defined as an option. Before transmitting, a station sends a RTS frame and if the receiving station is not expecting a frame from anyone else, it replies with a CTS frame. Upon receiving the CTS, the station that issued the RTS may begin transmission. The RTS-CTS frame contains information about how long it takes to transmit the data frame. When other stations within range of the receiving station hear the CTS, they refrain from transmitting and set a timer, called the network allocation vector (NAV). The NAV indicates the amount of time that must elapse before the current transmission is complete and the channel can be sampled for idle status. Figure 3.5 illustrates this operation.

An optional mechanism defined by the standard is point coordination function (PCF) to support time-bounded service. It is used only in infrastructure configuration. It provides a contention-free transmission through a point coordinator, which is a function performed by an AP. It is a master-slave approach, where the point coordinator is a polling master that coordinates all stations and determines which one can access the medium. Different QoS levels are supported through the use of a priority-polling mechanism. PCF has a higher priority than DCF and it may start transmission after a duration PCF interframe space (PIFS), where PIFS is less than DIFS and PIFS is greater than SIFS. A contention-free period (CFP) alternates with a contention period (CP) over time. A CFP and CP form a superframe. PCF is used to access the medium during CFP, and DCF is used during CP.

The point coordinator transmits a beacon frame to mark the beginning of a superframe. The beacon is transmitted at regular intervals, so each station knows when the next beacon frame will arrive, which is the target beacon transition time (TBTT) and is announced in every beacon. All stations set their NAV before the CFP. The point coordinator polls a station with data to transmit. If the station has pending data, it replies with an acknowledgment that can also be piggybacked with other data frames. If the point coordinator does not receive a reply after a period of PIFS, it polls the next station. If the point coordinator has pending data for the sta-

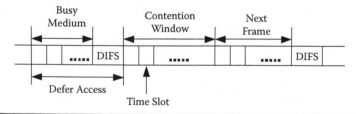

Figure 3.4 DCF operation.

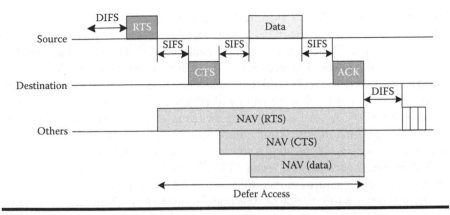

Figure 3.5   RTS/CTS operation.

tion, it combines the data and poll frames in a CF-poll frame. The polling continues until the CFP expires. The point coordinator terminates CFP by broadcasting a CF-end frame. Figure 3.6 illustrates the PCF operation.

3.2   IEEE 802.11b Standard (Wi-Fi)

In 1999, the 802.11 standard was revised and a new standard, 802.11b was released. This standard is more popularly known as Wi-Fi. It ensures compatibility with legacy 802.11 products by using the DSSS physical layer and operating at the same 2.4-GHz frequency band. It offers a maximum bit rate of 11 Mbps (comparable to a fixed Ethernet) with fallback rates of 5, 2, and 1 Mbps. This approach provides for smooth transition to faster WLAN systems. The access protocol used is similar to 802.11.

A Wi-Fi device typically consumes about 100 to 350 mA of power. The power management strategy specifies that a Wi-Fi device may be in either an awake or a doze state. In the doze state, the station cannot transmit nor receive. There are two power management modes: active mode (AM) and power save mode (PS). A station that wishes to go into PS must inform the AP by setting the power management

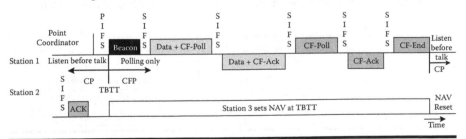

Figure 3.6   PCF operation.

bit in the packet header. In an infrastructured network, the AP stores all traffic addressed to stations that are in PS. When the AP transmits the periodic beacon, it lists the stations in PS and whether traffic is pending for them. Stations in PS switch to AM at regular intervals to receive the beacon. If there is pending traffic for them, the stations receive the traffic and then return to PS. In an ad hoc network, traffic destined for stations in PS is stored in a distributed fashion among active stations because there is no AP. All stations in PS switch to awake state in a temporal window, called an announcement traffic indication message (ATIM) window, during which stations that store pending traffic for other nodes send special frames called ATIM frames. If a station receives an ATIM frame, it remains awake to receive the traffic, otherwise it returns to PS until the next ATIM window.

Even though Wi-Fi quickly became popular due to its high bit rate, interference in the 2.4-GHz ISM band is a serious problem. To overcome this problem, IEEE released the 802.11a standard that operates in the 5-GHz frequency band.

3.3   IEEE 802.11a Standard

Because 802.11a operates at a different frequency band than 802.11 and 802.11b, it is incompatible with the previous two standards. Moreover, 802.11a uses orthogonal frequency division multiplexing (OFDM) to offer a higher bit rate of up to 54 Mbps. OFDM makes more efficient of the spectrum, and the high data rate is achieved by using up to 52 carriers to transmit data from a single source. Other speeds offered are 6, 12, and 24 Mbps. The major drawback in using the 5-GHz band is it is not license-free in all countries. To address this problem, IEEE enhanced the 802.11a standard by releasing the 802.11g standard.

The 802.11a standard defines a variable length packet. The MAC layer provides two types of service: asynchronous and contention-free. Like 802.11, the asynchronous service implements CSMA/CA with a binary exponential backoff known as DCF. It defines a basic access method and an optional four-way handshaking known as the RTS-CTS method.

Figure 3.7 shows the format of the physical protocol data unit (PPDU). The header contains the payload length, the transmission rate, a parity bit, and a tail. The rate field specifies the type of modulation and the coding rate used in the rest

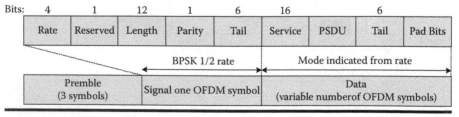

Figure 3.7   802.11a PPDU format.

of the packet. The length field specifies the number of bytes in the physical service data unit (PSDU) and takes a value between 1 and 4095. The parity bit is a positive parity for the first 17 bits of the header. The tail bits are used to reset the convolutional encoder and terminate the code trellis in the decoder. The first seven bits in the service field are set to zero and are used to initialize the descrambler, and the remaining bits are reserved for future use. The pad bits ensure that the number of bits in the PPDU maps to a number of OFDM symbols.

3.4   IEEE 802.11g Standard

The 802.11g standard offers a maximum bit rate of 54 Mbps and operates in the 2.4-GHz ISM band like 802.11, but uses OFDM instead of DSSS. It is compatible with 802.11b and provides a smooth transition from 802.11b for organizations that need to migrate to a WLAN offering a higher bit rate. In addition to OFDM, 802.11g supports the packet binary convolution code that offers bit rates of 22 and 33 Mbps as options. Its operating range is 100 m under ideal indoor conditions and 100 m outdoors.

3.5   HIPERLAN/2

HIPERLAN/2 is a European standard that operates in the 5-GHz band offering a maximum bit rate of 54 Mbps and makes use of OFDM. The spectrum allocation is 455 MHz in Europe, 300 MHz in the United States, and 100 MHz in Japan. The system can be deployed indoors and outdoors. In a normal operation, a MS communicates with other stations via the AP. For direct links (DiLs), stations in the same radio cell may exchange data directly. In both cases, the AP controls access to the medium and the assignment of radio resources to the stations. It provides support for multimedia applications.

3.5.1 The Convergence Layer

Figure 3.8 shows the HIPERLAN/2 protocol stack. The convergence layer performs two functions:

1. Adapts service requests from higher layers to the service offered by the data link control.
2. Converts the variable-size higher layer packet to a fixed-size service data unit (SDU) that is used within the data link control (DLC) layer.

This layer provides interworking functions to support different core networks, including the mapping of QoS parameters and traffic requirements, segmentation and reassembly (SAR), and header translation.

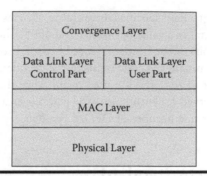

Figure 3.8 HIPERLAN/2 protocol stack.

3.5.2 The DLC Layer

The DLC layer is divided into two parts, namely the DLC control part and the DLC user part. It handles:

■ Association control: Association establishment, authentication, encryption key exchange, disassociation
■ DLC control: Connection setup, release, and modify, multicast
■ Radio resource control (RRC): Link adaptation, power control, dynamic frequency selection (DFS), handover

After an association, the station can request a dedicated control channel that is used to set up the radio bearers. A station may request multiple DLC connections, each with a unique QoS support that is determined by the AP.

The DFS operation enables an AP to select a frequency automatically based on interference measurements, accessible range of carriers, and a target carrier-to-interference ratio. The control part handles the exchange of data between an AP and a MS. The radio link quality may change with time, depending on the radio environment; for example, the traffic in surrounding cells. A link adaptation scheme is used to cope with the variation in both uplink and downlink channels. The physical layer mode is adapted based on the current link quality. The AP measures the link quality on the uplink channel and indicates in the frame control channel (FCH) which physical layer mode the station should use. Likewise, the MS measures the link quality on the downlink channel and suggests a physical mode to use in its resource request (RR) message, but the AP is responsible for the final selection of the physical mode to be used for both downlink and uplink.

The error control function detects and recovers transmission errors. The user part handles error control with three modes of operation:

1. Acknowledged mode provides retransmission that is based on selective repeat automatic repeat requests (ARQs) to ensure reliability.
2. Repetition mode offers some reliability and is based on repeating the data-bearing DLC protocol data units (PDUs). The transmitter can arbitrarily retransmit PDUs to achieve a more reliable reception. The receiver only accepts PDUs with sequence numbers (SNs) within the receiver's acceptance window.
3. Unacknowledged mode provides unreliable, low latency transmission. The receiver delivers all correctly received PDUs to the convergence layer. There is no feedback channel.

DLC is also responsible for traffic management. The mechanisms defined are:

- Wireless connection admission control: Responsible for accepting a new DLC connection based on QoS parameters, traffic requirements, and radio link quality.
- Wireless congestion control (WCC): Responsible for detecting congestion in the AP area cell. The MS and AP WCC take appropriate action when congestion is detected.
- Scheduler: Responsible for allocating capacity to peer DLC control traffic and DLC connections' traffic links. Capacity allocation is performed using TDMA time division duplex (TDMA/TDD). It also preserves the traffic characteristics on established DLC connections.

3.5.3 The MAC Layer

The medium access control (MAC) layer is designed to provide the QoS to support multimedia and real-time applications. Its medium access uses the TDMA/TDD approach with a MAC frame with a period of 2 ms. The standard defines a fixed length packet. The MAC scheme is based on a central controller located at the AP. The central controller determines the direction of information flow between the AP and the station at a given time. The flow utilizes a time slotted structure called a MAC frame (Figure 3.9), consisting of three time slots and three phases. The time slots are:

1. Broadcast control channel: Contains control information sent in every MAC frame to enable RRC functions.
2. Frame control channel (FCH): Contains an exact description of resource allocation within the current MAC frame.
3. Access feedback control channel (ACH): Conveys information on previous random access attempts.

The three phases are uplink to the AP, downlink from the AP, and DiL between two stations. The phases are allocated dynamically depending on the demand for

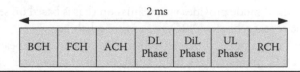

Figure 3.9 **HIPERLAN/2 MAC format.**

transmission resources. The allocation is handled centrally by the AP. Before transmitting, a station has to request capacity from the AP using the RACH.

A station with data to transmit requests capacity by sending a RR containing the number of pending long transport channel (LCH) PDUs that the station currently has for the particular DLC connection to the AP. The station may use the contention slots, which are based on slotted ALOHA, to send the RR message. If a collision occurs, the station is informed in the ACH of the next MAC frame, resulting in the station executing a backoff procedure. After sending the RR, the station goes into a contention-free mode where the AP schedules the downlink and uplink channels for the station. The scheduling is handled centrally by the AP, enabling efficient QoS support. From time to time, the AP may poll the station for more information about pending PDUs currently at the station.

There are two types of PDUs for the three phases:

1. A long PDU is 54 bytes long and contains control or user data with 48 bytes allocated for the payload. The remaining bytes are used for PDU type, a SN, and cyclic redundancy check (CRC). A long PDU is referred to as a LCH.
2. A short PDU is nine bytes long and contains only control data. It may contain RRs, ARQ messages, or other control information. A short PDU is referred to as a short transport channel (SCH).

Traffic to or from a station can be multiplexed onto one PDU train containing long and short PDUs (Figure 3.10). Each connection contains 54 octets LCH for data and 9 octets SCH for control messages.

Each user is assigned zero, one, or more slots in a frame. The number of slots assigned to a user varies from frame to frame depending on the amount of bandwidth requested by the station. The uplink and downlink packets are sent on the same frequency channel using TDD.

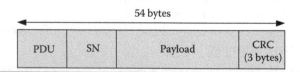

Figure 3.10 **HIPERLAN/2 PDU format.**

3.5.4 The Physical Layer

The physical (PHY) layer uses OFDM with a carrier spacing of 20 MHz. Table 3.1 lists the seven modes defined in this layer. The first six modes are mandatory; mode 7 is optional. The AP uses the link adaptation operation to select the highest possible physical layer mode for uplink and downlink channels based on radio link quality measurements. There are 52 subcarriers per channel: 48 subcarriers carry data and 4 subcarriers are pilots that facilitate coherent modulation. This layer maps the physical layer SDU received from DLC into a framing format, namely PHY-PDU, for transmission.

Functions performed by the physical layer are:

- Radio channel quality measurement and interrogation to determine the channel quality in terms of received signal strength and noise or interference level. The measurements are used to support dynamic frequency channel selection, handover, and link adaptation.
- Radio channel signaling rate interrogation.
- Physical layer initialization and configuration, for example, frequency range, RF power limits, coding schemes.
- Support and execution of radio channel selection.
- Antenna selection.
- Handover procedure.
- RF power settings per channel to allow the station to preset the RF power output of the transmitter on a per channel basis.

Table 3.2 compares the WLAN standards. ETSI BRAN (broadband radio access network), ARIB MMAC, and IEEE 802.11 cooperate closely to ensure that

Table 3.1 Physical layer modes.

Physical Layer Mode	Modulation and Code Rate	Data Rate (Mbps)
1	bpsk 1/2	6
2	bpsk 3/4	9
3	qpsk 1/2	12
4	qpsk 3./4	18
5	16 qam 9/16	27
6	16 qam 3/4	36
7	64 qam 3/4	54

BPSK—binary phase shift keying, QAM—quadrature amplitude modulation, QPSK—quadrature phase shift keying.

the PHY layers of the various 5-GHz WLAN standards are harmonized, which helps in facilitating low-cost production of devices conforming to all standards.

3.6 IEEE 802.1x Standard

802.1x is a port-based network access control standard that provides mutual authentication of a network and a client. It addresses the problem of rogue APs that can be easily installed in a corporate environment. The users authenticate themselves to the actual LAN, not the AP, through an authentication database such as remote authentication dial-in user service (RADIUS) server.

This standard utilizes an existing protocol—Extensible Authentication Protocol (EAP). Three components involved in the authentication are the supplicant (usually the client agent), the authenticator (usually the AP), and the authentication server (RADIUS). The user authentication data is transported using EAP and are encapsulated in 802.1x messages.

3.7 IEEE 802.11i Standard

Due to the ease of cracking the Wired Equivalent Privacy (WEP) protocol encryption, IEEE defines two stronger encryption algorithms in the 802.11i standard, namely WEP2 (an enhanced version of WEP) and advanced encryption standard (AES). It uses the Temporal Key Integrity Protocol (TKIP) as a patch for legacy 802.11 to correct vulnerabilities in the WEP protocol, particularly the reuse of encryption keys. It adds a 64-bit message integrity check, an extended initialization vector with packet sequencing, a rekeying mechanism, and per-packet key mixing in addition to a 128-bit data encryption. AES lets administrators specify a 128-, 192- or 256-bit key. When AES is ratified as part of the revised WEP standard, it eliminates the use of the 24-bit initialization vector (IV), which is the main weakness in the current version of WEP. Although AES may address the current security

Table 3.2   A comparison of WLAN standards.

Standard	Bit Rate (Mbps)	Spectrum (GHz)	Transmission	Compatible with
802.11	2	2.4	FHSS/DSSS	802.11b, 802.11g
802.11a	54	5	OFDM	None
802.11b	11	2.4	DSSS	802.11
802.11g	54	2.4	OFDM	802.11, 802.11b
HIPERLAN/2	54	5	OFDM	None

concerns, its deployment will not be problem-free. Vendors have to replace existing APs and other equipment to comply with the new standard. The use of a longer key means that client devices will need extra processing power to perform encryption and decryption, possibly causing a performance slow down. The additional processing requirement results in more power consumption on WLAN cards.

TKIP, a temporary solution until 802.11i is ratified, uses a 128-bit temporary key, but all users on a specific AP share the same key, which means that if one user is compromised, everybody is vulnerable to an attack. The difference between TKIP and WEP is that TKIP changes the key every 10,000 packets, whereas WEP keys are static.

3.8   IEEE 802.11e Standard

There are a few drawbacks of the 802.11 PCF that make it difficult to provide QoS guarantee:

- When a point coordinator schedules the beacon transmission at TBTT, the beacon can only be transmitted when the medium has been idle for at least PIFS, resulting in a possible delay in transmitting the beacon frame.
- The delay in transmitting the beacon results in delays in the transmission of time-bounded packets that have to be delivered during CFP. There is a possibility that delivery may not be finished before the next TBTT. The unpredictable delays in each CFP may severely affect QoS guarantees.
- Unknown transmission duration of a polled station because the station is allowed to send a frame of arbitrary length (maximum of 2304 bytes), thus, affecting the QoS guarantee to other stations.

802.11e is a supplemental standard to provide QoS guarantees in a WLAN. It defines enhancements to the 802.11 MAC with the introduction of enhanced distributed coordination function (EDCF) and hybrid coordination function (HCF). A station that is 802.11e-compliant is termed an enhanced station. An enhanced station that acts as a centralized controller for all other stations within the same QBSS (a BSS that contains 802.11e-compliant HC and stations) is termed a hybrid coordinator (HC). A HC is a function in a 802.11e-compliant AP. There are still two alternating phases in a superframe: CP and CFP. EDCF is used during CP only when HCF is used in both phases.

3.8.1 EDCF

Traffic is classified into traffic categories (TCs). A station may generate more than one type of traffic and each type of traffic is handled independently of the others.

Frames are delivered according to multiple backoff instances at a station, where each backoff instance is based on a specific TC parameter. During CP, a station contends for a transmission opportunity (TXOP) and independently starts a backoff after detecting an idle channel for a period of arbitration interframe space (AIFS), where AIFS is greater than PIFS and AIFS is greater than or equal to DIFS. The value of AIFS may be increased for each TC. After AIFS expires, each backoff sets a counter to a random value from the interval [*I*, *CW*+1]. The minimum size is (*CW* min[*TC*]), where *CW* is a TC-dependent parameter. If the channel is detected to be busy before the counter reaches zero, the backoff has to wait for the channel to be idle before restarting AIFS. The backoff counter is reduced by one at the beginning of the last slot interval of AIFS when the channel is idle during AIFS; whereas in legacy DCF, the counter is reduced at the beginning of the first slot after DIFS.

An unsuccessful transmission causes the value for *CW* to be increased based on the persistence factor *PF*[*TC*], that is, *newCW*[*TC*] ≥ {(*oldCW*[*TC*] + 1) × *PF*} -1, and another backoff counter is calculated based on the new *CW* to reduce the probability of another collision. This is unlike legacy 802.11 where the value of *CW* is doubled after an unsuccessful transmission.

Each station may have up to eight transmission queues, each with its own QoS parameters that determine its priority (Figure 3.11). If more than one counter reaches zero at the same time, a scheduler in the station avoids a virtual collision by granting TXOP to the TC with the highest priority. TXOP is the duration a station has the right to transmit, starting at start time up to a maximum duration. There are two types of TXOP:

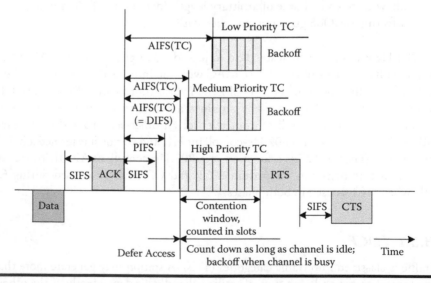

Figure 3.11   802.11e multiple backoff for traffic with different priorities.

1. EDCF-TXOP: Granted by contention and the duration is constrained by a QBSS-wide limit distributed in the beacon frame.
2. Polled-TXOP: The duration is specified by the duration field inside the poll frame.

The prioritized channel access is governed by the QoS parameters per TC, namely *AIFS[TC]*, *Cw* min[*TC*], *OF[TC]*, and *CW* max[*TC*] (optional). The parameters may be adapted over time and are announced periodically in the beacon frames.

3.8.2 HCF

HC may only grant a TXOP to itself or another station after the channel has been idle for PIFS. During CP, a TXOP begins when either (AIFS plus backoff time) has expired or the station receives a poll frame, QoS CF-Poll, from the HC. HC may send a QoS CF-Poll after PIFS without any backoff and may issue polled TXOPs during CP using its prioritized medium access.

During CFP, stations do not contend for the channel. Only the HC can grant TXOPs using the QoS CF-Poll frame. The start time and maximum duration of a TXOP is specified using the QoS CF-Poll frame. The CFP ends after the time announced in the beacon frame or when HC transmits a CF-end frame.

The 802.11e standard defines a random access protocol for fast collision resolution. This mechanism allows stations to request for polled TXOPs by sending RRs without contending with other DCF/EDCF traffic. A controlled contention occurs during a controlled contention interval that starts when the HC sends a specific control frame. The control frame forces legacy stations to set their NAV and remain silent during the interval. The control frame defines controlled contention opportunities (i.e., short intervals separated by SIFS) and a filtering mask containing the TCs in which RRs may be placed. A station that has pending traffic with a TC that matches the filtering mask chooses an opportunity interval and transmits a RR frame containing the requested TC and TXOP duration or the queue size of the requested TC. For fast collision resolution, the HC acknowledges the reception of a request by generating a control frame with a feedback field so that the requesting station can detect collisions during controlled contention.

3.9   Security Issues

The 802.11 standard defines the WEP to encrypt messages using a Rivest Cipher 4 (RC4) pseudorandom number generator with two key structures: 40 and 128 bits. RC4 is a symmetric algorithm where a secret key is deployed at the AP and the station. The secret key is used for encrypting data and checking data integrity. It is also used by an AP to authenticate stations. The goals of WEP are:

■ Confidentiality to prevent eavesdropping.
■ Access control to protect access to the wireless network infrastructure. There is an option to discard all packets not encrypted using WEP.
■ Data integrity to prevent tampering with the transmitted message, which is achieved using an integrity checksum.

Even though the standard allows the key to be up to 104 bits long, the implementation usually uses a 40-bit key, which is too short. WEP concatenates the 40-bit secret key with a 24-bit IV to produce a 64-bit RC4 key stream. Because the IV is sent to the receiver in plain text so that the receiver can generate the same key stream, an attacker can see the first 24 bits of each key sent using WEP. The standard allows the IV to be reused, so an attacker can easily collect the IV and use it to deduce the secret key shared by the AP and the station.

Another drawback of WEP is its poor key management, where the keys in a device may be left unchanged for long periods. If the device is lost or stolen, an attacker can use the key to compromise the device and any other devices that share the key. Dynamic key management solutions could help address this problem.

WEP's implementation of the CRC-32 algorithm calculates a 32-bit checksum to verify the integrity of packets sent over the air. Because the checksum is not encrypted, a side-channel attack may compromise data integrity.

An AP has to be configured properly before deployment in a WLAN. Most manufacturers ship APs with WEP disabled with default parameter values that must be changed. Among the parameters that have to be configured are the AP's password, simple network management protocol parameters, channel selection, Dynamic Host Configuration Protocol (DHCP) setup, and the integrated firewall configuration. If it is not configured properly, an attacker can easily compromise the network by utilizing the default values. For example, because products from the same vendor have the same default service set identifier (SSID) that is required to access the network, it is imperative that the SSID is changed. At the same time, it is also critical to turn off the AP's broadcast mode that broadcasts the unit's SSID. Otherwise, anyone with a wireless network interface card (NIC) could get the SSID from the AP itself, which renders changing the SSID pointless.

A device's MAC address is its unique ID and can be used to authenticate a device. Unfortunately, because the MAC address is broadcast in plain text, an attacker can easily eavesdrop on a transmission to obtain the MAC addresses of authorized devices, and later, use the address to gain unauthorized access to the network. If a NIC with the MAC address defined as "allowed" is stolen, the attacker can use the card to gain access to the WLAN. If the theft is reported, the network administrator has to update the MAC address filter on all APs to prevent unauthorized access, which might be time-consuming on a large network. An enterprise wireless gateway (EWG) reduces the hassle of updating this information by providing a centralized point of security.

Instead of using brute force to decrypt a message, an attacker can use Internet Protocol (IP) redirection to trick an AP into decrypting a message for it. An attacker sniffs an encrypted packet transmitted over the air and modifies it so that it is sent to a new destination address that the attacker controls. When the modified packet arrives at the AP, it decrypts the packet and sends the decrypted packet to the new destination where the attacker can read it.

Another technique to trick an AP is a reaction attack that exploits a Transmission Control Protocol (TCP) characteristic that requires an acknowledgment to be sent for each packet that is received error-free. An attacker can identify an acknowledgment packet by its size—the attacker does not need to decrypt a packet to ascertain if it is an acknowledgment. An attack is launched by intercepting a ciphertext, flipping a few bits in the ciphertext, and adjusting the CRC accordingly to obtain a new ciphertext with a valid WEP checksum. The modified ciphertext is transmitted to the AP. The attacker then waits to see if the eventual recipient sends a TCP ACK, which tells the attacker whether the modified text passes the TCP checksum and was accepted. By carefully choosing the bits to flip, the attacker ensures that the TCP checksum remains undisturbed. Thus, the presence or absence of an ACK packet reveals one bit of information about the unknown plaintext. The attacker can then deduce the remaining unknown bits using classical techniques.

To gain access to the WLAN, an attacker has to gain access via an AP. Therefore, it is important that an AP should be placed at a secure point to avoid easy access to it, thus, compromising security. A particular threat is what is termed as a rogue AP, which is an AP installed by an unauthorized employee or an attacker. Employees (with or without malicious intent) may install a rogue AP from their cubicles to connect to the corporate network without the knowledge of the network administrator. If the AP is not properly configured, it may become a security hole in the network. Even though a standard specifies the range that signals can travel (i.e., the coverage area), it is not unusual for signals to travel beyond the specified coverage area (Figure 3.12). An attacker does not have to be in the building to install a rogue AP. An attacker may sit outside the building and gain access to the corporate network through an AP whose signal has traveled beyond the corporate network boundary. This type of attack is known as "war driving." Signals may travel beyond the corporate office boundary and leak into parking lots, another company's office, and public places. This problem can be solved by walking around the company's perimeter with a spectrum analyzer to check whether signals travel beyond the desired area. This also helps in detecting rogue APs. If the signal is found to travel beyond the desired coverage area, the network administrator can take corrective actions by lowering the AP's signal strength, moving it to a location where its signal does not go beyond the desired area, or using different types of antennas to control signal strength and direction. Smart antennas that can control an AP's signal to confine it to a specific area can also help mitigate this problem.

Hot spots not only can be used to gain access to the Internet, but also to private and corporate networks. You may use the hot spot at the airport to access the

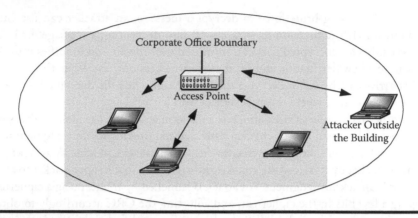

Figure 3.12   Corporate office boundary versus WLAN coverage area.

corporate server while waiting for your flight. By doing so, you open a channel to your corporate network. Because public networks, such as hot spots, provide minimal security, if any, attackers can sniff the network and view all packets transferred in plaintext. This problem can be avoided by configuring the system to disallow access from public networks but doing so would limit the user's mobility and service availability. Another option is to use a virtual private network to secure connection through public WLANs by creating a tunnel or using a secure encrypted connection between the mobile user and the destination. Installing a personal firewall on the user's device also provides added protection.

Steps that can be taken to prevent unauthorized access include:

- Use EWG to authenticate users and filter unauthorized users.
- Use a centralized key distribution server to easily and effectively manage WEP keys. However, this gives rise to the possibility of a single point of failure.
- Use an intrusion detection system to monitor the WLAN.

3.10   IP over 802.11 WLAN

A few APs can be interconnected through an IP routed network to form the WLAN IP network. Referring to Figure 3.13, an access router (AR) connects one or more APs to the network. The APs exchange IP packets with the ARs. They also act as ARP proxies for the MSs associated to them. At any one time, a MS is connected with one AR, termed a serving AR. When a MS moves across APs connected through the same AR, an intra-AR handover takes place, and an inter-AR handover takes place when the APs are connected to different ARs. An intra-AP handover is handled using Inter-AP Protocol. For inter-AP handover, the new AR interacts with the MS to perform an IP handover.

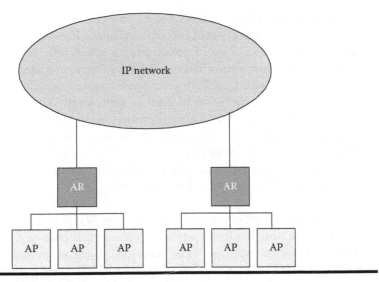

Figure 3.13 A WLAN IP network.

3.11 Integrating 802.11 WLAN and UMTS

There are two possible scenarios for the deployment of WLANs in public areas (Jaseemuddin 2003). It may be a hot spot that is deployed as a standalone network that is not connected to a UMTS network. You may subscribe to UMTS and WLAN services through the same service provider; in which case, it is cost-effective for the operator to use a common authentication and billing infrastructure. The other possibility is a roaming scenario where the WLAN operates in conjunction with a UMTS network, where the WLAN is used to support high-speed data transfer. UMTS on its own is able to sustain a per user data rate of a few hundred kilobits per second and is limited by the total cell capacity of up to 2–3 Mbps. Given that a WLAN is able to support a data rate of up to 54 Mbps, it is used to augment the packet data service of UMTS. Users may use a dual-mode station to access the two networks, where one interface connects to the WLAN, and the other connects to UMTS. The networks are configured such that WLANs form microcells within UMTS macrocells.

The following issues have to be addressed before integrating UMTS and 802.11 WLANs:

■ Assess the impact in QoS differences of the two networks on the types of applications that users can run and, hence, the type of traffic handled by each network. UMTS support four levels of QoS—interactive, voice, stream, and best-effort—whereas currently, 802.11 only supports best-effort traffic until the implementation of PCF provides support for isochronous traffic.

■ Be aware that GPRS and WLAN support different connection paradigms, that is, connection-oriented and connectionless, respectively.

■ Decide on the handling of packet routing across two networks with different mobility management schemes. GPRS packets are routed through tunnels established between GPRS gateway support node–serving gateway support node GGSN-SGSN and SGSN–radio network controller (RNC), whereas packet routing in IP networks may be achieved by using tunnels (e.g., Hierarchical Mobile IP version 6 [HMIPv6]) or host-specific routing (e.g., cellular IP).

■ Decide whether to connect a WLAN to RNC, SGSN, or GGSN. The best integration point has to be selected taking into consideration different cost-performance benefits for different scenarios.

The integration architecture in Figure 3.14 uses SGSN as the integration point, not RNC or GGSN. The RNC performs radio-specific tasks (e.g., converts packets into radio frames, manages radio resources). Because the radio interfaces of WLAN and RNC are totally different, making RNC an integration point would require major revision of complex radio procedures. Hence, it is not an option. Using GGSN as an integration point is also unsuitable because during handover to UMTS, SGSN has to re-create the mobility state and acquire or reestablish the Packet Data Protocol (PDP) session and radio access bearer (RAB) context. Because this information is not held at GGSN, making it the integration point would slow down the handover process.

This architecture is designed for use in hot spots and is designed to fulfill the following requirements:

■ Users in a WLAN use 802.11 to establish a data connection and the UMTS radio network subsystem (RNS) to establish a voice connection. The connections can be established in parallel.

■ WLAN is used to augment the data service, thus, allowing users to run applications that they cannot run if they are connected only to UMTS.

■ Users in UMTS network use UMTS RNS for both data and voice services.

When a MS moves to a WLAN microcell, the packet data connection through UMTS RNS is dismantled and reestablished through the WLAN, but a voice connection is still maintained through UMTS RNS. For this to be possible, the user must use a dual-mode station with two interfaces that allows parallel access to the WLAN and UMTS. In this architecture, the SGSN is the integration point and it sees the WLAN and UMTS as two different networks.

Referring to Figure 3.14, the WLAN is connected to SGSN through border routers. A connection through a UMTS network requires explicit signaling between the MS and the network to establish and manage the bearer path. The MS in a WLAN cell maintains connectivity to both networks through different inter-

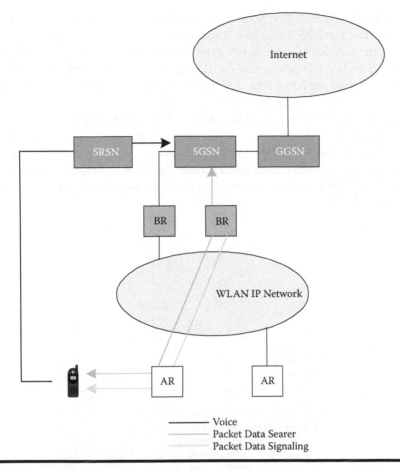

Figure 3.14   WLAN-UMTS integration architecture. (From Jaseemuddin, M. 2003. An architecture for integrating UMTS and 802.11 WLAN networks. *Proceedings of the 8th IEEE International Symposium on Computers and Communications (ISCC '03)* **2:716–723. Used with permission.)**

faces. Both bearer and signaling paths for PS connections are established through WLAN to the SGSN.

If a node is switched on in a UMTS cell, it only receives beacons from a UMTS BS, thus, it activates the UMTS interface and executes the UMTS power-up procedure. If it is switched on in a WLAN, it can be connected to both UMTS and WLAN. It receives beacons from both the UMTS BS and 802.11 AP. First, it runs the UMTS-GPRS power-up procedure through the UMTS interface during which it ignores the beacons received from the 802.11 interface. On power-up, it attaches to GPRS and establishes the basic data service (a best-effort bearer service) connection through serving RNS (SRNS) with the SGSN and GGSN. The PDP context is established at the GGSN, SGSN, and the MS, whereas the mobility context is established at the SGSN and MS. The SRNS performs RAB setup for the best-effort bearer service. After performing the UMTS power-up procedure, it responds to the 802.11 beacons by running the association procedure with the AP. Once it is associated with the AP, it runs the UMTS-WLAN handover procedure to handover the basic data connection to WLAN (Figure 3.15).

The GGSN assigns an IP address for the MS, which it uses to communicate with other nodes in the Internet. This IP address is not guaranteed to be topologically correct for the WLAN the MS is associated to. Therefore, it may request for a temporary topologically correct address from the WLAN, called a care of address (COA). The COA is used as the tunnel endpoint for packets tunneled between the SGSN and the MS.

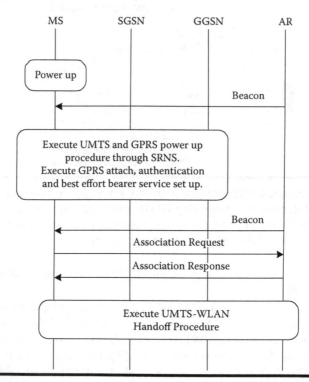

Figure 3.15 Power up procedure. (From Jaseemuddin, M. 2003. An architecture for integrating UMTS and 802.11 WLAN networks. *Proceedings of the 8th IEEE International Symposium on Computers and Communications (ISCC '03)* **2:716–723. Used with permission.)**

3.12 Summary

The WLAN standards are specified in IEEE 802.11. There are two types of configuration: infrastructure mode with at least one central AP connected to the wired network and an ad hoc or peer-to-peer mode. It defines three RF physical layers that operate at the 2.4-GHz ISM band: FHSS, DSSS, and IR. Instead of CSMA/CD, 802.11 specifies DCF, also known as CSMA/CA to overcome the hidden station problem.

The 802.11b standard, more popularly known as Wi-Fi, offers a maximum bit rate of 11 Mbps (comparable to a fixed Ethernet) with fallback rates of 5, 2, and 1 Mbps. The access protocol used is similar to 802.11. Even though Wi-Fi is very popular, interference in the 2.4-GHz ISM band is a serious problem. To overcome it this problem, IEEE released the 802.11a standard that operates in the 5-GHz frequency band.

Another WLAN standard is HIPERLAN/2, a European standard that operates in the 5-GHz band offering a maximum bit rate of 54 Mbps and makes use of OFDM. It provides support for multimedia operations.

Current trends show that WLAN will coexist with traditional wired LANs to provide wider coverage to users and also to support anytime, anywhere access to services and information. The integration of WLANs and LANs gives rise to a number of security concerns, some of which have been adequately addressed, but others remain a topic of active research.

Reference

Jaseemuddin, M. 2003. An architecture for integrating UMTS and 802.11 WLAN networks. *Proceedings of the 8th IEEE International Symposium on Computers and Communications (ISCC '03)* 2:716–723.

Bibliography

Borisov, N., I. Goldberg, and D. Wagner. 2001. Intercepting mobile communications: The insecurity of 802.11. *Proceedings of the 7th Annual International Conference on Mobile Computing and Networking (Mobicom)*:180–188.

Demestichas, P., A. Oikonomou, G. Vivier, and M. Theologou. 2003. Management of wireless home networking technologies in the context of composite radio environments. *Mobile Computing and Communications Review* 7(3):33.

Doufexi, A., S. Armour, M. Butler, A. Nix, D. Bull, J. McGeehan, and P. Karlsson. 2002. A comparison of the HIPERLAN/2 and IEEE 802.11a wireless LAN standards. *IEEE Communications Magazine* 40(5):172.

Khun-Jush, J., G. Malmgren, P. Schramm, and J. Torsner. 2000. Overview and performance of HIPERLAN type 2—A standard for broadband wireless communications. *Proceedings of the IEEE 51st Vehicular Technology Conference (VTC 2000* 1:112–117.

Mangold, S., S. Choi, P. May, O. Klein, G. Hiertz, and L. Stibor. 2003. IEEE 802.11e wireless LAN for quality of service. *IEEE Wireless Communications* 10(6):40–50.

Park, J. S., and D. Dicoi. 2003. WLAN security: Current and future. *IEEE Internet Computing* 7(5):60.

Prasad, N. R., and H. Teunissen. 1999. A state-of-the-art of HIPERLAN/2. *Proceedings of the IEEE 50th Vehicular Technology Conference (VTC 1999)* 5,2661–2668.

Varshney, U. 2003. The status and future of 802.11-based WLANs. *Computer* 36(6):102.

Yeh, J. H., J. C. Chen, and C. C. Lee. 2003. WLAN standards. *IEEE Potentials* 22(4):16.

Online Resources

IEEE 802 Standards. http://www.ieee802.org/11/ (Accessed February 5, 2007).

HIPERLAN Standard. http://portal.etsi.org/radio/hiperlan/hiperlan.asp (Accessed February 5, 2007).

Chapter 4

Wireless Personal Area Networks

A wireless personal area network (WPAN) focuses on personal operating space around a person or object that typically extends within a 10-m radius. IEEE classifies three classes of WPANs based on data rate, battery drain, and QoS.

1. High-data-rate WPAN: Defined in the 802.15.3 standard, is suitable for multimedia applications that requires high QoS.
2. Medium-rate WPAN: Defined in the 802.15.1 (Bluetooth) standard, handles various tasks such as cell phones and PDA communications and supports QoS for voice applications.
3. Low-rate WPAN (LR-WPAN): Defined in the 802.15.4 standard, is targeted at applications with lower power consumption and cost requirements not considered in other WPAN classes. LR-WPAN is characterized by a low data rate, on the order of a few hundred kilobits. The 802.15.4 standard is discussed in detail in Chapter 5.

In addition to the above standards, the HomeRF Working Group (HRFWG) defines the HomeRF standard that is also meant for WPAN.

4.1 HomeRF

The consortium of companies that formed the HRFWG was disbanded in 2003 and has ceased to develop, promote, or distribute the HomeRF specification. However, the specification is discussed briefly to compare it with other WPAN technologies and show the evolution of the technology.

Figure 4.1 shows the HomeRF protocol stack that defines the MAC and physical layers to support wireless voice and data networking in a residential area. The Shared Wireless Access Protocol (SWAP) is the product of cooperation between the industry and HRFWG to develop a protocol for radio-based home networks. The specification defines a common air interface to support wireless voice and LAN data services in the home. It provides higher data rates and ensures interoperability between different vendor products.

SWAP operates in the 2.4-GHz band and combines elements of digital enhanced cordless telecommunication (DECT) and the IEEE 802.11 standards. TDMA is used to deliver time-critical services, and CSMA/CA is used to deliver high-speed data. Because it closely resembles the IEEE 802.11 MAC and PHY layers, and is a subset of DECT standard, the SWAP MAC layer can support both data-oriented service and voice service.

A SWAP system can operate as either:

■ An ad hoc network that supports data communications only. It is a peer-to-peer network where all stations are equal and control of the network is distributed between stations.

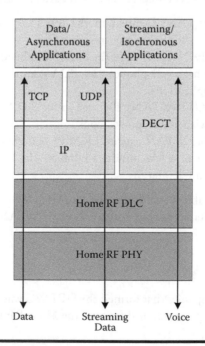

Figure 4.1   HomeRF protocol stack.

A managed network under the control of a connection point or AP. It handles time-critical services, for example, voice applications. A connection point coordinates the system and provides a gateway to the PSTN. The connection point is also used to support power management to extend battery lifetime by scheduling device wake-up and polling.

The network can support up to 127 nodes that may be one of the following types:

A connection point that supports voice and data services

A voice terminal that uses TDMA to communicate with a BS

A data node that uses CSMA/CA to communicate with a BS and other data nodes

Voice and data nodes that use both types of service

4.2   Bluetooth Technology

Bluetooth, conceived in 1994 by Ericsson Mobile Communications, is a personal area network (PAN) technology that has received a lot of attention. The name Bluetooth comes from the Danish King Harald Blåtand (Bluetooth in English) who united the Scandinavian countries for the first time. Bluetooth technology has brought together major players in the telecommunication industries, such as Ericsson, Intel, Toshiba, and Nokia to name just a few.

Bluetooth technology aims to be rid of cables by providing wireless connections between communication devices—for example, a PDA, a hand phone, and a laptop—so that you can transfer data between these devices seamlessly without the hassle of connecting them with cables. The aims of the technology are:

To provide low power radio chip that can be installed in any device without causing power drain.

To offer low-priced Bluetooth chips that are affordable for most people—the targeted price is US$5.

To have a small footprint because it is intended to be embedded in small devices.

To enable both speech and data transmission, preferably simultaneously.

4.2.1 Basic Operation

Bluetooth operates in the unlicensed 2.4-GHz ISM band (2402–2480 Hz). Because this is the same frequency band for WLANs and microwave ovens, the technology has to be robust in handling interference. The channel access technique used is

FHSS. A data rate of 1 Mbps is achieved using Gaussian-shaped frequency shift keying (GFSK) modulation. The spectrum is divided into 79 channels with each channel allocated a bandwidth of 1 MHz. It supports both point-to-point and point-to-multipoint connections.

Devices within communication range of each other form a piconet. Communications in a piconet are controlled by a device that acts as a master. Other devices act as slaves. At any one time, there can only be one master device and seven active slaves in a piconet. There can be up to 255 slaves in park mode that do not exchange data but synchronize with the master. Two or more overlapping piconets form a scatternet. A device can be a member of more than one piconet but can act as master in only one piconet. Figure 4.2 shows two piconets that overlap to form a scatternet. A scatternet enables a multihop wireless network where two nodes can communicate with one another even though they are not directly connected by using intermediate nodes as relays.

When a device is powered on, it may operate as a master or slave device. First, it has to listen for a master's inquiry for new devices and respond to it. This phase allows the master to know the address of the slave device. It also allows the device to learn the master's address and the clock phase that is used to compute the frequency hopping sequence that is unique to the piconet. The master determines the frequency hopping sequence (based on its Bluetooth address) and the phase of the hopping sequence (based on its clock). Once the slave's address is known, the master may open a connection with the slave provided the slave is listening for paging requests. If the slave responds to the paging request, the master and slave synchronize over the frequency hopping sequence. After a connection is established, the devices may authenticate each other before they start communicating. The connection may be torn down after communication is completed, and the device may enter one of the power-saving modes (PSMs).

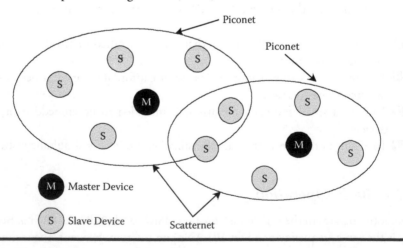

Figure 4.2 Two piconets and a scatternet.

A Bluetooth device has a range of 10 m. Bluetooth supports the allocation of up to 20 piconets, each with a maximum aggregate data transfer rate of 400 kbps.

4.2.2 Bluetooth Protocol Stack

The Bluetooth protocol stack allows devices to locate and connect to others to facilitate data exchange and execute interoperable, interactive applications. The protocol stack consists of three components (Figure 4.3):

1. Transport protocol group: This group consists of the Logical Link Control Adaptation Layer Protocol (L2CAP), link manager, baseband, radio, and host controller interface (HCI) layers (Figure 4.4). It provides a means for Bluetooth devices to locate each other. It supports the creation, configuration, and management of physical and logical links so that higher layer protocols and applications can pass data through the transport protocols.
2. Middleware protocol group: This group consists of the RF communications (RFCOMM), Service Discovery Protocol (SDP), telephony control specification (TCS), and audio layers (Figure 4.4). It utilizes the transport protocols and presents standard interfaces to the application layer for communication across the transports. It includes third-party and industry standard protocols (e.g., IPs, wireless application protocols, Infrared Data Association) in addition to protocols defined specifically for Bluetooth. Three protocols are designed specifically for Bluetooth: RFCOMM (provides support for legacy applications that interface with a serial port to operate seamlessly over Bluetooth transport protocols), TCS (for advanced telephony operations), and mobility support for cordless handsets and BSs.
3. Application group: This consists of applications that utilize Bluetooth links. These may be legacy applications unaware of Bluetooth transports or Bluetooth-aware applications.

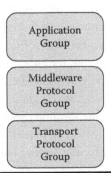

Figure 4.3   Bluetooth protocol stack components.

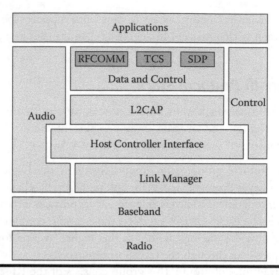

Figure 4.4   Bluetooth protocol stack.

Figure 4.4 shows the Bluetooth protocol stack. Each layer in the protocol stack performs a set of defined functions:

- Radio: The Bluetooth radio layer defines parameters to optimize over-the-air packet transmissions for short-range wireless communications.
- Baseband: Provides the functionality for air interface packet framing and the establishment and maintenance of piconets and link control. It defines the master and slave roles, how devices find each other, and the establishments of connections. It also determines the frequency hopping sequences and how the air interface is shared between synchronous and asynchronous traffic.
- Link manager: Responsible for link setup and control. Also handles authentication, encryption, QoS, transmission scheduling, and physical parameters control.
- HCI: Provides a mechanism for the higher layers to delegate the decision on whether to accept connections to the link manager. Also decides whether to switch on filters at the link manager. The radio, baseband, and link manager may be packaged into one module. The HCI layer defines a common interface to access the module to allow interoperability between modules by different vendors. HCI offers a single standard interface for the upper layers to access the lower layers. The module uses HCI commands to enter certain modes of operation, for example, an authentication operation.
- Control: Relays control information between layers. For example, L2CAP notifies the link manager of its expected QoS or an application may send control information to set the power-saving mode for the device.

■ L2CAP: Provides connection-oriented and connectionless data services. It provides functions to perform protocol multiplexing, SAR, and QoS support. Version 1.2 of the Bluetooth specification allows IP to be implemented over L2CAP. The master-slave concept does not propagate to higher layers. From L2CAP and above, communication is based on a peer-to-peer model, and there is no provision for different actions for a master or slave device.

■ RFCOMM: Emulates serial ports. Because most serial communication involves a cable for transferring data across serial ports, Bluetooth has to provide support for serial communications for the initial cable-replacement usage model. For example, legacy applications that benefit from this feature are peer-to-peer file transfer and data synchronization. RFCOMM provides a virtual serial port to facilitate smooth migration of applications modeled for cabled serial communications to wireless communications.

■ SDP: Unlike wired networks where services are provided by servers through static configurations, devices forming an ad hoc network cannot rely on servers to provide the services they require. SDP allows Bluetooth devices to discover what services are available in neighboring devices. SDP defines a standard method for Bluetooth devices to discover each other's presence and the services offered by one another. For example, a Bluetooth-enabled camera uses SDP to detect whether there is a digital printer nearby.

■ TCS and audio: TCS provides telephony functions such as call control and group management. TCS is used to set up the call parameters. Once a call is established, the audio channel carries the voice content. TCS is compatible with the Q.931 specification. It is suitable for mobile office workers and small office or home office environments.

4.2.3 Frequency Hopping

If you want to listen to the broadcast of your favorite radio station, you have to tune to the frequency at which it is broadcast. Radio stations always broadcast at the same frequency. When we transmit data over the air, we can also take the same approach. However, this is not always a good idea because of interference and security concerns. Broadcasting at the same frequency all the time makes it easier for eavesdroppers to listen to transmissions. Using the frequency hopping technique, the frequency at which data is transmitted is changed periodically. An advantage of frequency hopping is that if interference occurs at a certain frequency band, only a very small number of packets will be affected and the corrupted packets are retransmitted later. Because the packet size is small, frequency hopping is more robust to interference. Figure 4.5 illustrates how frequency hopping works. Channel 1 (C1) hops from F1 to F4 to F2 to F3 to F1. If interference occurs on F2, only one packet might be lost.

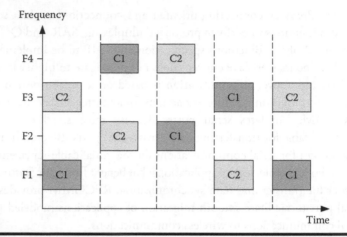

Figure 4.5   Frequency hopping.

A Bluetooth device changes its transmission frequency at a rate of 1600 times per second. The transmission frequency remains unchanged for 625-µs long slots. Each slot is assigned a sequence number. The master transmits in even slots, and a slave transmits in odd slots. A message may last for one, three, or five consecutive slots. Another advantage of frequency hopping is it provides a degree of security because only a receiver that knows the frequency hopping pattern can receive and assemble all packets of a message.

4.2.4 The Bluetooth Protocol

Bluetooth defines two types of links: asynchronous connectionless link (ACL) and synchronous connection-oriented link (SCO). An SCO packet is fixed-length and fits into one slot with a fixed 64 kbps throughput. The packet length can be 10, 20, or 30 bytes. An ACL packet fits into one, three, or five slots. The payload length is from 17 to 339 bytes with symmetric throughput ranging from 108.8–433.9 kbps. Its asymmetric throughput is 732.2/57.6 kbps.

A SCO link provides guaranteed delay and bandwidth, not withstanding Link Manager Protocol messages that have higher priority. A slave may establish a maximum of three SCO links with the same master or two SCO links with different masters. On the other hand, a master may establish a maximum of three SCO links with three different slaves. The SCO's 64-kbps constant bit rate symmetric channel is suitable for streaming applications such as average quality voice and music. It provides limited reliability. There is no retransmission nor is CRC applied to the payload. Forward error correction (FEC) is offered as an option.

ACL provides support for non-real-time traffic. Only one ACL link may be established between a slave and the master, meaning that it cannot support applications with different QoS parameters. The master determines a schedule that is used by a slave to exchange packets, one at a time, with the master. ACL links may be symmetric or asymmetric with different preset bandwidths. A 16-bit CRC is applied to the payload for error protection. Optional features offered are 2/3 FEC convolutional code and ARQ. ACL links are configured via an interface offered by the link manager. Parameters that can be configured are:

Type of QoS: None, best effort, and guaranteed best effort (default).
Token rate: The data transfer rate guaranteed on that link.
Token bucket size: The buffer size for the received data. The default is zero.
Peak bandwidth.
Latency.
Delay variation: The maximum allowable difference between packet delays.

A slave specifies the value of the parameters in a request to the master that is handled by the admission control function implemented by the master's link manager. If the master accepts the QoS request, it configures the link with the slave by setting two parameters:

1. Poll interval: The maximum time interval between two consecutive transmissions
2. N_{BC}: The number of retransmissions for broadcast packets

Violation of the QoS parameters may be communicated to the upper layers by the link manager.

4.2.5 Bluetooth Security

All Bluetooth headers are protected by an 8-bit CRC. Datagram payloads on ACL links are protected by a 16-bit CRC, but stream payloads on SCO links are not protected. A Bluetooth packet may be protected by 1/3 FEC, 2/3 FEC, or not at all.
Bluetooth specifies three security modes:

Mode 1: Not secure
Mode 2: Service level enforced security (after channel establishment)
Mode 3: Link level enforced security (before channel establishment)

Authentication and encryption are handled by the following entities:

The unique 48-byte Bluetooth device address assigned to a device
■ A private authentication key (random number)
A private encryption key (random number)

■ A 128-bit frequently changing random number, dynamically generated by each device

There are two security levels: trusted and nontrusted. There are three levels of service: open services, services requiring authentication, and services requiring authentication and authorization. A personal ID number (PIN) code that is between 1 and 16 octets must be entered at initialization of each communicating device. The PIN may also be hardwired in the device.

Communicating devices authenticate each other using a shared secret termed as a link key. The link key is established during a session called pairing during which the link key is computed using the address of each device, a random number, and a shared secret (PIN). If both parties need to be authenticated, the pairing process is executed twice—once for each party. The secret key can be entered manually when the device is used for the first time or hardwired for paired devices that are always used together. Data encryption is performed using the E0 stream cipher. A unique encryption key is generated for each session. Per-packet keys are derived from the encryption key.

4.2.6 Power Management

Bluetooth is designed specifically for portable devices with short range and limited battery power. On average, a Bluetooth device consumes between 1–35 mA of power, which is much lower compared to the 100–350 mA typically consumed by Wi-Fi devices. When a Bluetooth device is not connected, it is in standby mode. When a device is connected, it operates in one of the following modes:

■ AM: The device is active and listens to transmissions from the master. It receives packets to help it to remain synchronized with the master and informs it when it can transmit to the master. This state provides the fastest response time but also consumes the most power.

■ Sniff mode: This is a low-power consumption mode where the device listens only during sniff slots. A slave becomes active periodically. When a device is in sniff mode, the master transmits packets for that slave only at certain regular intervals, called sniff intervals. The slave listens for packets from the master only at the start of each interval. If it receives a packet at the start of the interval, it continues listening to receive the packet. Otherwise, it sleeps until the next interval. The power consumption and responsiveness of a slave in sniff mode depends on the length of the sniff interval.

■ Hold mode: A slave stops listening for packets from the master for a specified time interval, called a hold time, that is agreed to by the master and the slave. During the hold time, the device may establish links to other devices or it may sleep. It resumes listening to the master at the end of the hold time. This

mode allows greater power savings than sniff mode but may be less responsive than sniff mode, depending on the hold time.
- Park mode: The device is not active but remains synchronized with the master. Because there may be more than seven slaves in a piconet, the park mode allows the master to orchestrate communications in a piconet by exchanging active and parked slaves to maintain up to seven active connections. A parked slave maintains synchronization with the master by listening to the master periodically using a beaconing scheme. This mode consumes the least power and is also the least responsive.

4.2.7 Bluetooth Usage Models

The usage models describe possible Bluetooth applications. The cordless computer usage model removes cables from computers, keyboards, mice, speakers, printers, scanners, and other computer peripherals. In addition to doing away with the hassles of dealing with cabling, it also offers freedom in placement and uses. Devices such as a keyboard or a mouse can be moved at the user's convenience without being constrained by the cable length. A printer can be shared without connecting it to a wired network: Move your laptop or the printer so that they are within range of each other, and you can print your document.

The headset usage model defines a headset that can be used not only with mobile phones but also with standard telephones. The latter is particularly useful, for example, at help desks and reservation offices where the receptionist needs to keep her hands free to use a computer. Because the headset is designed based on a standard specification, it can be used with multiple devices that adhere to the standard; for example, it can be used with a cordless telephone and for audio interaction with computers.

Nowadays, many people use three types of telephones: a telephone at home, another one at the office, and a mobile phone. Would it not be more convenient if we could use one telephone for all purposes? This is what the three-in-one phone usage model tries to accomplish. The three-in-one phone allows a mobile phone to be used as a cellular phone, as a cordless phone, and as an intercom or walkie-talkie. Let us say you are out shopping and you need to make a phone call. Your mobile phone is able to detect your surroundings and establish a connection in the usual manner, and your service provider charges the normal call rate using a mobile phone. When you return home, your mobile phone is able to detect that you are home and acts as a cordless phone. Any calls you make are charged the standard landline call charge. When you are at the office, your mobile phone detects your surroundings and allows you to use the phone as a walkie-talkie to talk to a colleague. In this case, no call charge is incurred.

The file transfer usage model allows temporary links to be set up quickly between devices to exchange files and other data objects such as business cards. It provides support for quick data exchange at meetings and conferences without the hassle of setting up and configuring a network among the devices.

The Internet bridge usage model provides two methods for using Bluetooth as a bridge to establish networks. The first method connects a computer to the Internet using a telephone to dial to an Internet Service Provider minus the cables. Assuming the computer and the telephone support the dial-up networking profile, a connection to the Internet is achieved wirelessly. A traveling user can connect to the Internet without having to wait to return to the hotel to plug in his modem to dial up to the corporate network. Figure 4.6a illustrates the dial-up networking usage model. The second method is direct network access and is typically used in enterprises and similar environments. Direct network access is achieved via a Bluetooth data AP that allows Bluetooth devices to connect to it wirelessly (Figure 4.6b). The data AP is connected to a LAN. This profile allows the LAN to be extended using a data AP without requiring any cabling between the AP and the devices.

A user is likely to use a number of devices during a workday. The PC is used to access information while at the office. The PDA is used to access the corporate database while meeting a customer. The mobile phone is used for a quick discussion with the boss. While meeting with the customer, the user may update the calendar and add a task to the to-do list using the PDA. Back at the office, the user updates the calendar and to-do list on the PC manually. The automatic synchronizer usage model allows easier management of personal data such as calendar entries and to-do lists so that the user's data on the PDA can be automatically synchronized with the data on the PC. The synchronizer merges data from two sources based on a set of rules such that the resulting datasets reflect identical information. The synchro-

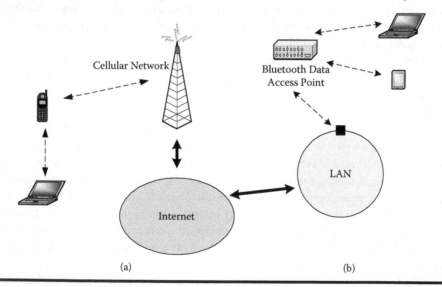

(a)   (b)

Figure 4.6   Internet bridge usage model.

nization is performed automatically without the user's intervention and is enabled by proximity networking, which is when two devices automatically synchronize whenever they come within range of each other.

The instant postcard usage model allows a photo taken with a digital camera to be sent to a laptop or mobile phone and later e-mailed to friends and family. This is useful not only for personal use (e.g., sending holiday photos), but also for commercial use (e.g., sending a photo of an antique grandfather clock to a potential customer).

The usage models discussed above give an overview of what is possible with Bluetooth technology. Applications that would be enabled by this technology, however, are not limited to the ones described above. What is possible is limited only by the creativity of the designers.

4.3   IEEE 802.15.3 Standard

IEEE 802.15.3 is designed to support ad hoc networking and multimedia QoS guarantees, where a node may join or leave a group or subnetwork, and plays the role of a master or slave node. The physical layer is similar to the one defined for 802.11b. It operates in the 2.4-GHz band at 11 Mbaud. It is designed to achieve data rates between 11 to 55 Mbps to support high-definition video and high-fidelity audio. Table 4.1 shows the modulation format and data rates defined in the 802.15.3 standard. The signals occupy a bandwidth of 15 MHz, allowing for up to four fixed channels.

The superframe structure is shown in Figure 4.7. The beacon contains network-specific parameters, such as information for new devices to join the network and power management. The beacon is followed by time slots of the contention access period (CAP) that makes use of CSMA/CA. CAP is for the transmission of frames that do not require a QoS guarantee. CAP is followed by time slots allocated under the guaranteed time slots (GTS) period. GTS is used for the transmission of image files, standard and high-definition video (e.g., Motion Picture Experts Group formats— MPEG1, MPEG2), and high quality audio.

The 802.15.3 standard is optimized for short-range transmission—a maximum distance of 10 m. It keeps costs low and allows integration into small consumer devices. The physical layer requires low power consumption, less than 80 mA, while actively receiving or transmitting data.

Table 4.1   Modulation format and data rate.

Modulation	Data Rate (Mbps)
Trellis code QPSK	11
Uncoded QPSK	22
16-/32-/64-QAM	33/34/55
QAM—quadrature amplitude modulation, QPSK—quadrature phase shift keying.	

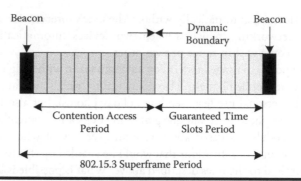

Figure 4.7   IEEE 802.15.3 superframe structure.

4.4   Home Area Networks

Activities in a home are different from activities in offices (Meyer and Rakotonirainy 2003). Office activities are more formal, structured, and task-oriented and aim to optimize productivity. Activities at home are less structured—people are free to decide how to organize their time and activities and with whom they interact. The social interactions that happen at the office also differ from the ones at home. A system that is useful in an enterprise environment is not necessarily suitable at home.

In an enterprise environment, an investment is only made on systems that deliver cost savings or productivity gains. The system should be easily installed and maintained and should not require users to have sound technical knowledge to keep it up and running. A system in an enterprise environment must take the necessary measures to protect data and intellectual property from competitors. Thus, security is a critical feature that must be incorporated into the system. On the other hand, designing a context-aware system for homes presents a different set of challenges. A home area network should give real personal value and must fulfill usability, social acceptance, privacy protection, low-cost and zero administration requirements. Zero administration is important because, unlike an office environment, home users do not have skilled technical personnel to assist them with setting up or configuring the system.

To be useful, a context-aware home has to be aware of its occupants' contexts, desires, whereabouts, activities, needs, emotions, and situations. This awareness enables it to adapt its interaction with the occupants. Context is measurable and relevant information that effect an occupant's behavior. Context information must be gathered as unobtrusively as possible and with the least effort from the occupants. Therefore, information is gathered mostly by sensors and not via user input.

A context-aware home is made of the following components:

■ Instrumentation: The building blocks such as sensors, wireless networks, and UIs that gather context information and allow human-computer interaction.

■ Middleware: The infrastructure that gathers context information, process it, and derive meaningful actions from it. A hardware abstraction layer provides decoupling from the actual implementation of sensors and wireless networks. A context manager processes the context information. A privacy manager handles privacy issues.

■ Applications: Use the gathered context information to infer what is expected by an occupant and delivers the service.

■ User experience: Everything felt, observed, and learned through awareness and interaction. A good experience meets the occupant's expectation, causes minimal frustration, and is enjoyable.

■ Privacy: Information gathered about an occupant's preferences and habits should be guarded against theft and unauthorized access. The security and privacy aspect must be considered during the initial stage of designing the context-aware home.

Figure 4.8 shows the interactions between the components and an occupant.

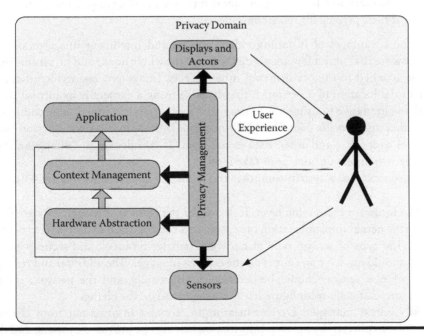

Figure 4.8   The components of a context-aware home interacting with an occupant. (From Meyer, S., and A. Rakotonirainy. 2003. A survey of research on context-aware homes. *Proceedings of the Australasian Information Security Workshop Conference on ACSW Frontiers* **21, 59. Used with permission from the Australian Computer Society Inc.)**

4.4.1 Instruments for a Context-Aware Home

Because most context information should be gathered with very minimal input from the user, sensors are used extensively in a context-aware home. The types of sensors used may vary because it is unlikely that one type of sensor will be suitable for an application. It is likely that data from multiple sensors is combined to increase the meaningfulness of the derived context and to improve error detection and correction. Sensors may be attached to objects or people to gather context information. Alternatively, unobtrusive video cameras may also be used to monitor the occupants, objects, and their environment.

4.4.2 Choosing Sensors for a Context-Aware Home

There are three main approaches to protecting privacy:

1. Try to minimize the amount of information that flows out of a system.
2. Restrict the amount of information acquired and stored to a minimum.
3. Give users more control and awareness over the collection and use of personal information so that they can choose if they are willing to give up certain parts of their privacy for more convenience.

The advantages of installing video cameras and intelligent image-processing systems are that only a few are needed to cover the whole house and no sensors need to be attached to the occupants or other objects. Image-processing identifies and tracks the location of a person. A drawback of using a camera is its intrusiveness, and an alternative to using it is to use smart mobile devices or wearable computers to gather and process contextual information. This approach gives a person more control over what and when data is collected. If you decide to "disappear," you simply switch off your device or take it off.

A context-aware system comprises the following components (Figure 4.8):

- ■ Hardware abstraction layer: It decouples the context manager software from the actual implementation of sensors and the communication infrastructure. The type of sensor, type of network, transfer protocol, and security policy should not have any affect on the context manager. The addition and removal of new sensors should be detected automatically, and the network should automatically reconfigure itself to accommodate the change.
- ■ Context manager: Derives meaningful context information from the raw data collected by sensors. Information can also be provided by users, and this is termed profiled information.
- ■ Privacy manager: Users should be able to choose which privacy information they want to gain more convenience from. They should also be able to grant or reject rights to other family members, guests, or smart objects in the home.

To gain users' acceptance, a context-aware home should be pleasant, personalized, and easy to use, and it must offer solutions to real-life problems. One of the appealing aspects of a smart home is the time savings gained from the ability to control devices in the house. Usability of software and UI is critical. Usability is defined as the effectiveness, efficiency, and satisfaction that users find in achieving specific goals in a particular environment.

4.5   Summary

Bluetooth is the most popular WPAN standard. It operates in the unlicensed 2.4-GHz ISM band. The channel access technique used is FHSS. A data rate of 1 Mbps is achieved using GFSK modulation. Initially, a Bluetooth device had a range of 10 m, but this has now increased to 100 m. Devices within communication range of each other form a piconet. Overlapping piconets form a scatternet.

Another WPAN standard is IEEE 802.15.3, which is designed to support ad hoc networking and multimedia QoS guarantees. A node may join or leave a group or subnetwork and play the role of a master or slave node. It operates in the 2.4-GHz band at 11 Mbaud and is designed to achieve data rates between 11 and 55 Mbps to support high-definition video and high-fidelity audio.

Use of WPANs are not restricted to offices—it can also be used to build home networks. Even though the same technology can be used at an office and in a home, the way it is used is markedly different because activities in a home are different from activities in an office. Office activities are more formal, structured, and task-oriented, and they aim to optimize productivity; activities at home are less structured—people are free to decide how to organize their time, activities, and interactions. The social interactions that happen at the office also differ from the ones at home. Even though WPAN may be used to share information in the office and at home, a home network is more likely to be used for entertainment. Regardless of how WPAN technology is utilized, security and privacy remain important concerns.

Reference

Meyer, S., and A. Rakotonirainy. 2003. A survey of research on context-aware homes. *Proceedings of the Australasian Information Security Workshop Conference on ACSW Frontiers* 21, 59.

Bibliography

Ferro, E., and F. Potortì. 2005. Bluetooth and Wi-Fi wireless protocols: A survey and a comparison. *IEEE Wireless Communications* 12(1):12.

Gutierrez, J. A., M. Naeve, E. Callaway, M. Bourgeois, V. Mitter, and B. Heile. 2001. IEEE 802.15.4: A developing standard for low-power low-cost wireless personal area networks. *IEEE Network* 15(5):12.

Miller, B. A., and C. Bisdikian. 2001. *Bluetooth revealed: The insider's guide to an open specification for global wireless communications.* 2nd ed.. Indianapolis, IN: Prentice Hall PTR.

Vaxevanakis, K., T. Zahariadis, and N. Vogiatzis. 2003. A review on wireless home network technologies. *Mobile Computing and Communications Review* 7(2):59.

Online Resources

Bluetooth official Web site. http://www.bluetooth.com/ (Accessed February 7, 2007).

IEEE 802.15 Working Group for WPAN. http://grouper.ieee.org/groups/802/15/ (Accessed February 7, 2007).

Home Area Networks Project. http://www.cl.cam.ac.uk/Research/SRG/HAN/second.html (Accessed February 7, 2007).

Chapter 5

Wireless Sensor Networks

A wireless sensor network (WSN) is a collection of sensor nodes that are distributed in a selected geographical area to monitor certain aspects of the environment. It incorporates technologies from three areas: sensing, communications, and computing. A sensor node performs the following functions:

- Senses environmental physical parameters, for example, temperature, vibration, humidity
- Processing of raw data locally to extract characteristic features of interest
- Temporary storage of the data
- Uses the wireless link to transmit data to its neighbors

The usage of WSN can be classified into three categories: monitoring space (e.g., environmental and habitat monitoring, precision agriculture, indoor climate control, intelligent alarms), monitoring objects (e.g., structural monitoring, ecophysiology, medical diagnostic, urban terrain mapping), and monitoring interaction between objects with each other and the encompassing space (e.g., wildlife habitats, disaster management, emergency response). A sensor node is also referred to as a mote or a probe.

This chapter discusses how sensor networks are used for various purposes and the architecture and protocols involved in building a WSN. It starts with an overview of WSN applications.

5.1 Applications of WSNs

WSN technology enables monitoring in remote or dangerous locations that were rarely studied due to inaccessibility. Before designing and installing a WSN for environmental monitoring, an understanding of its physical environment is imperative, for example, the system must be able to withstand conditions such as extreme temperature or vibration.

WSN allows the monitoring of a habitat without requiring heavy and bulky equipment. For example, Culler et al. (2004) conducted a study for environmental monitoring in collaboration with a biologist. Instead of hauling a suite of instruments weighing 13.5 kg up a tree with cables dangling down to the forest floor, an entire wireless weather station fits in a tube about the size of a film canister. Sixteen sensor nodes were placed at four elevations in a 35-m redwood tree. The sensors monitor humidity, barometric pressure, and temperature. For example, the sensors collect data every five minutes and compute an average temperature at each elevation. The measurements show that there is a climatic weather variation between the top of the tree and the forest floor. The variation of several factors, such as photosynthesis rate, water transport rate, and the scale of respiration, influence the microclimate around a tree and, in turn, influence the habitat dynamics of species in and on the tree.

The GlacsWeb project (Martinez et al. 2004) monitors a glacial environment. The monitoring of ice caps and glaciers provides information about global warming and climate change. It focuses on how the ground on which glaciers rest affects their movement. Nine probes are embedded at the ice-sediment boundary at the depth of 50–80 m. The probes must be nonintrusive and mimic the movement of stones and sediment under the ice. A probe is composed of temperature and tilt sensors, a snow meter, a Webcam, and a differential GPS (dGPS) to follow ice movement. The probes are scheduled to wake up every four hours to take readings, and a real-time clock (RTC) powers off between readings.

Motion monitoring involves the monitoring of physical structures, such as machines, airplanes, and bridges, where variations in behavior indicates wear, fatigue, or other mechanical changes. For example, vibration sensors are attached to various pieces of a machine, and every few months, a team of electricians tours the plant and collects readings from the sensors. The data is sent to a central computer that analyze it for signs of wear. Appropriate steps can then be taken to repair or replace components in the machine based on the analysis of the data. Alternatively, instead of the electrician team touring the plant, the WSN can be configured to send the collected data to a central computer periodically.

The examples given above show that sensor nodes are used in different environmental surroundings. Sensor nodes must be designed to withstand the possibly harsh environment they are embedded in. In addition, there are a few other common requirements for sensor nodes.

5.2   Requirements for WSNs

Because WSNs are often deployed in remote areas that are not easily accessible, sensor nodes should be:

Low cost because a large number may be required to monitor an environment
Unobtrusive but durable under a range of environmental stresses
Energy efficient so that they can remain in situ with little human interaction
Robust to withstand errors and failures
Low polluting
Maintenance-free for years at a time
Able to reliably interconnect with a cyber-infrastructure that permits frequent network access for data upload, device programming, and management

Its communication requirements are omnidirectional high-power probes, long-range communication between BSs and the reference station, a low data rate, error detection and corrections, and backup channels. Traditional network abstractions are unsuitable for WSNs, for example, operating systems for such networks must tightly integrate wireless connectivity.

5.3   WSN Architecture

There are three types of architecture: infrastructure-based architecture, ad hoc architecture, and infrastructure-assisted architecture. In an infrastructure-based architecture, a sensor node registers with a BS and all communications are done via the BS. A node sends a packet destined for another node to the BS. If the destination node is registered with the BS, the BS forwards it to the destination node. Referring to Figure 5.1, node C sends a packet to node A via BS1. Otherwise, the packet is forwarded to another BS where the destination node is registered. For example, when node G sends a packet to node I, BS1 forwards the packet to BS2 who then delivers it to node I. The advantage of infrastructure-based architecture is that it is simple and there are no power restrictions on the BSs.

In an ad hoc architecture, there is no BS. If the destination node is within the range of the node, the packet is sent directly to the destination. Otherwise, the packet is relayed via intermediate nodes that forward it to the intended node. Referring to Figure 5.2, node C sends a packet to node F directly because they are within transmission range of each other. However, when node A sends a packet to node J, it has to be routed via node F and node I. This approach is termed multihop transmission. The advantages of this architecture are its versatility, lower power consumption, and easy, quick deployment in remote areas.

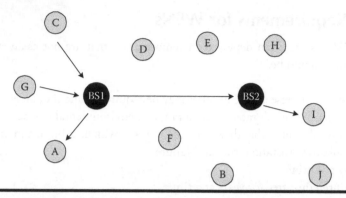

Figure 5.1   Infrastructure-based architecture.

An infrastructure-assisted architecture is a hybrid of the previous two architectures. A node outside the range of the BS could use other nodes to relay packets to the BS. A packet may or may not go through the BS depending on the location of the source and destination nodes. Referring to Figure 5.3, node A sends a packet to node F directly because they are within transmission range. Node G, which is not within transmission range of the BS, has to route its packet via node C. The BS then delivers the packet to its intended destination, node B. Two or more different infrastructure-based networks can exchange packets by relaying packets through the sensor nodes thus forming a single network. An example of this application are building sensors that relay information to BSs located in the building. The information can be made available to control rooms and remote locations, if necessary.

5.4   The 802.15.4 Standard

The 802.15.4 standard was developed for LR-WPAN. It is a global standard designed to address the needs of low data rates, low power consumption, and low-cost wire-

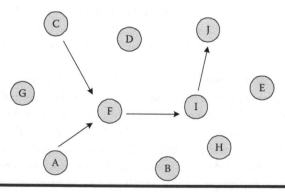

Figure 5.2   Ad hoc architecture.

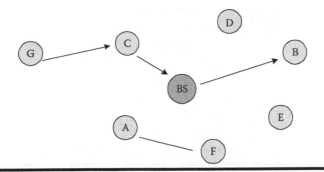

Figure 5.3   Infrastructure-assisted architecture.

less networking among inexpensive fixed, portable, and moving devices in residential and industrial environments. Why can existing standards, such as Bluetooth, not be used for this purpose? Bluetooth is unsuitable because to support a wider range of applications and provide QoS, it has become more complex. The 802.15.4 standard defines only 45 MAC primitives and 14 PHY primitives—one-third the total number defined in Bluetooth. Bluetooth's complexity makes it inappropriate for certain simple applications requiring low costs and low power consumption. Among applications that benefit from the 802.15.4 standard are:

- Automation and control in homes, factories, and warehouses
- Monitoring for public safety (e.g., floods), health, and environment (e.g., gas leaks)
- Entertainment (e.g., learning games and interactive toys)

The standard supports one-hop star and multihop peer-to-peer topologies. There are two types of devices: full function devices (FFDs) and reduced function devices (RFDs). An FFD is required to support all 45 MAC primitives and can function in any topology. It can talk to reduced function devices (RFDs) and FFDs. It operates in one of three modes: PAN coordinator, coordinator, or device. A RFD is required to support 38 MAC primitives under minimal configuration. Its use is limited to star topology. It can only talk to a FFD and is meant for very simple applications. It cannot act as a network coordinator. A network requires at least one FFD as a network coordinator. Endpoint devices may be RFDs to reduce cost.

The standard only defines the PHY layer and the data link layer (DLL) (Figure 5.4). The network and upper layers are addressed by ZigBee.

5.4.1 The DLL

The DLL is divided into two sublayers: MAC and logical link control (LLC). MAC provides support for association and disassociation, acknowledged frame delivery,

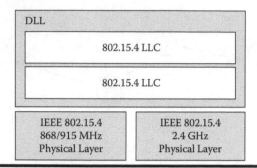

Figure 5.4   IEEE 802.15.4 layers.

channel access mechanism, frame validation, guaranteed time slot management, and beacon management. The association and disassociation functions allow for an automatic setup of a star network and the creation of a self-configuring peer-to-peer network. LLC is common among the 802 standards, such as 802.3 and 802.11.

The MAC frame (Figure 5.5) is known as the MAC PDU and consists of the MAC header, MAC SDU, and MAC footer. The MAC header specifies the MAC frame being transmitted, the format of the address field (8-bit or 64-bit address), and acknowledgment. The address field varies between 0 and 20 bytes because a data frame would contain source and destination information but an acknowledgment frame does not contain any address information. The maximum MAC frame size is 127 bytes.

There are four types of frames: beacon, data, acknowledgment, and command. The data and beacon frames contain information sent by the higher layers. The acknowledgment and command frames originate from the MAC layer and are used for peer-to-peer communications.

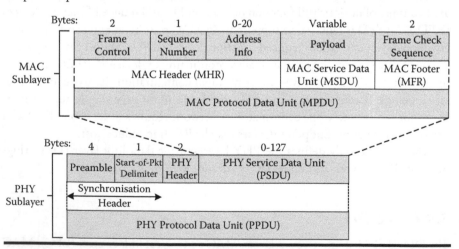

Figure 5.5   The 802.15.4 MAC and PHY frames format.

The frame check sequence field is used to verify the integrity of the data frame. A transaction is deemed successful when the sequence number in an acknowledgment frame matches the one in a previously transmitted data frame.

The network operates in either beacon-enabled mode or non-beacon-enabled mode (Figure 5.6). In beacon-enabled mode, a coordinator broadcasts beacons periodically to synchronize the attached devices to identify the PAN and to describe the structure of superframes. In non-beacon-enabled mode, a coordinator does not broadcast beacons periodically, but may unicast a beacon to a device that is soliciting beacons.

An optional superframe mode is defined to support low latency communication. A dedicated network coordinator, termed a PAN coordinator, transmits superframe beacons in predetermined intervals between 15 ms and 245 s. The duration of a superframe is announced to the attached network devices by the coordinator in its beacon. The time between two beacons is divided into 16 equal time slots that are independent of the duration of the superframe (Figure 5.7). A device that wishes to transmit during the CAP has to compete with other devices using CSMA/CA. Alternatively, the coordinator may assign time slots to a device that requires dedicated bandwidth or low latency transmission. The assigned time slots are known as GTS. The coordinator may allocate up to seven GTSs, and a GTS may occupy more that one slot period. The GTS consists of a CFP that always appears at the end of the active superframe starting at a slot boundary immediately following CAP. A sufficient portion of CAP remains for contention-based access and all contention-based transactions must be completed before CFP begins. A device transmitting in GTS must make sure that its transmission is complete before the next GTS or the end of the CFP.

An important function performed by MAC is acknowledging the successful reception of a frame. When a data or command frame has been successfully received and validated, an acknowledgment frame is returned immediately. A sender indicates whether it expects an acknowledgment using the frame control field.

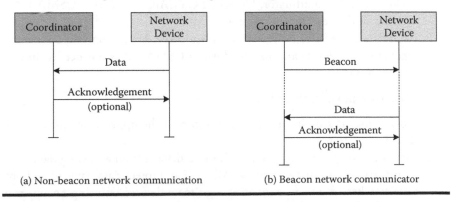

(a) Non-beacon network communication   (b) Beacon network communicator

Figure 5.6   Beacon-enabled and non-beacon-enabled communications.

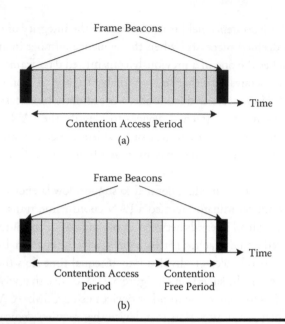

Figure 5.7   Superframe structure.

Data is transferred from a device to a coordinator, from a coordinator to a device, or from a device to a peer in a peer-to-peer topology. There are three types of data transfers:

1. Direct data transmission: Applies to all data transfers. Data transmission is handled using unslotted CSMA/CA or slotted CSMA/CA depending on whether non-beacon-enabled or beacon-enabled mode is used.
2. Indirect data transmission: Applies only to data transfers from a coordinator to its devices. The coordinator keeps a transaction list and waits for extraction by the corresponding device. A device checks the beacon frames for pending packets from its coordinator. Unslotted CSMA/CA or slotted CSMA/CA is used in the data extraction procedure.
3. Guaranteed time slot (GTS) data transmission: Applies only to data transfers between a device and its coordinator. CSMA-CA is not needed in data transmission.

The standard provides three levels of security:

1. No security: If security is not important, or if the upper layers provide sufficient protection.
2. Access control lists: Prevents unauthorized devices from accessing data.
3. Symmetric key security: Employs AES-128 to protect data payload and prevent attackers from impersonating legitimate devices by using an integrity code (IC).

The key distribution method is not specified in the standard to minimize the cost for devices, but may be included in the upper layers of the applications if needed.

5.4.2 The PHY Layer

There are 14 PHY primitives with 2 PHY options that are based on DSSS methods:

1. 2.4-GHz PHY operates in the 2.4-GHz ISM band with data rate of 250 kbps.
2. 868/915-MHz PHY operates in the 868-MHz band in Europe and 915-MHz ISM band in the United States with data rates of 20 and 40 kbps, respectively.

Table 5.1 summarizes the characteristics of the PHY layer. The 868/915-MHz PHY supports one channel in the 868-MHz band and ten channels in the 915-MHz band. The 2.4-GHz PHY supports 16 channels with a channel spacing of 5 MHz to ease transmitting and receiving filter requirements. The standard defines a dynamic channel selection but the specific selection algorithm is left to the network layer. A scan function in the MAC layer steps through the list of supported channels to search for a beacon. Low-level functions defined in the PHY layers perform receiver energy detection, link quality indication, and channel switching to enable channel assessment and frequency agility.

Both PHY layers have the same frame structure (Figure 5.5). The PPDU contains a synchronization header, a PHY header, and a PSDU. The synchronization header consists of a 32-bit preamble to attain symbol and chip timing. It may also be used for coarse frequency adjustment. Seven bits of the PHY header are used to specify the payload length in bytes. The maximum payload length is 127 bytes. A typical payload length for home appliances is expected to be between 30 and 60 bytes. The maximum packet duration is 4.25 ms for the 2.4-GHz band, 26.6 ms for the 915-MHz band, and 53.2 ms for the 868-MHz band.

The transmission range is between 10 and 20 m but may be extended with a moderate increase in transmit power. A star topology can provide complete home coverage. Devices operating in the 2.4-GHz band are vulnerable to interference by other services operating in the same band. However, this is not a critical issue because the standard is targeted at applications with relatively low QoS requirements that do not require isochronous communication, and that are expected to

Table 5.1 Summary of physical layer characteristics.			
Band	*Available Channels*	*Maximum Data Rate*	*Modulation Technique*
2.4 GHz	16	250 kbps	O-QPSK
915 MHz	10	40 kbps	BPSK
868 MHz	1	20 kbps	BPSK
BPSK—binary phase shift keying, O-QPSK—offset quadrature phase shift keying			

perform multiple retries occasionally for successful transmission. A critical requirement for the applications is long battery lifetime that is achieved by low transmitting power and very low duty cycle operation. The devices are expected to be in sleep mode 99.9 percent of the time.

5.4.3 Power-Saving Mechanisms

The 802.15.4 standard defines power-saving mechanisms that are based on the beacon-enabled mode. In direct data transmission, if the BatteryLifeExtension option is set to TRUE, the receiver of the beaconing coordinator is disabled after maxBattLifeExtPeriods (default value is 6) backoff periods following the interframe space (IFS) period of the beacon frame. Using the default configuration, the transceiver of a coordinator or device is required to be turned on for only 1/64 of the duration of a superframe if there is no data to be exchanged.

In indirect data transmission, a device may enter a low-power state upon finding no pending packets when checking the beacon received from its coordinator. GTS data transmission also has a low duty cycle but it is considered costly for low data rate applications. The small CSMA/CA backoff period and short transceiver warmup time also help to further reduce power consumption.

5.5   The ZigBee Protocol

ZigBee operates on top of the IEEE 802.15.4 standard to address issues in the upper protocol layers and application profiles by adding logical network, security, and application software. ZigBee is promoted by 70 companies, such as Honeywell, Mitsubishi, Motorola, Philips, and Samsung, that form the ZigBee Alliance. ZigBee was introduced because there was no standard that addressed the unique needs of remote and sensor network applications. Sensors and control devices do not need high bandwidth, but they need low latency and very low power consumption. There are various proprietary wireless systems, and the systems are often not interoperable. ZigBee Alliance offers a standardized base set of solutions to address these concerns. ZigBee offers product interoperability and vendor independence that result in a winwin situation for consumers. The alliance hopes to produce ZigBee chips at a low price of $2(US) because low cost is an important requirement for sensor nodes.

ZigBee defines modules for quality assessment, receiver energy detection, and clear channel assessment. CSMA is used to determine which node may transmit to avoid collisions. The maximum distance between nodes is up to 70 m. ZigBee supports three network topologies: star, mesh, and cluster tree (a hybrid of the first two). It uses a master-slave configuration suitable for star networks comprising many infrequently used devices that transmit small data packets. It supports up to 254 nodes and can be increased if necessary. Master nodes can talk to each other.

An important feature is that ZigBee incurs low latency. A node that is powered down can wake up and transmit a packet in about 15 ms. This feature is important for time-critical messages. For example, sensors operating in a nuclear plant need their messages transmitted immediately.

Operating on top of the 802.15.4 standard, a ZigBee network consists of nodes that may act as a coordinator (FFD), a router (FFD), or end device (FFD or RFD) as shown in Figure 5.8. A coordinator performs the following functions: sets up a network, transmits network beacons, manages network nodes, stores network node information, and routes messages between paired networks. A coordinator requires extra random-access memory (RAM) for the node database, transaction table, and pairing table. Typically, it operates in the receiving state. A network node searches for available networks, transfers data from its application or environment, determines whether there is data pending, and requests data from the network coordinator.

Figure 5.9 shows the ZigBee stack system. The network layer allows extensible coverage by allowing clusters to be added. Networks can be consolidated or split. This layer is responsible for:

- Establishing a new network
- Providing the capability to join and leave a network
- Configuring a new device to carry out the necessary operations
- Synchronizing among devices in the network through either tracking beacons or polling
- Applying security policy for outgoing frames and removing security to terminating frames
- Routing frames to their destinations

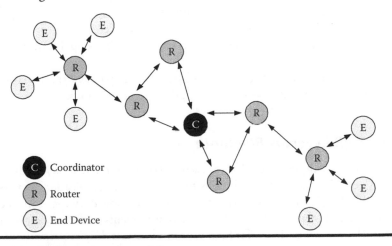

Figure 5.8   ZigBee network model.

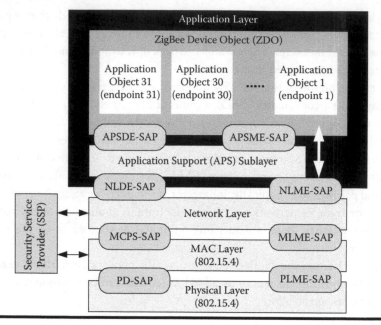

Figure 5.9 ZigBee stack system.

The routing algorithm used is a hierarchical routing strategy with table-driven optimization where possible. The algorithm is based on ad hoc on-demand distance vector (AODV) and the cluster-tree algorithm.

The application layer consists of:

■ Application support (APS) sublayer: Maintains the table for binding to match two devices based on their services and needs, discovery to find devices, and forwarding of messages between bound services.
■ ZigBee device object: Defines the role of a device, initiates or responds to binding requests, and establishes a secure relationship between network devices.
■ Manufacturer-defined application objects that implement the actual application according to ZigBee-defined application descriptions.

5.5.1 ZigBee versus Bluetooth

Why do we need ZigBee when there are already other standards, such as Bluetooth? Bluetooth and ZigBee are two solutions for two different application areas as illustrated in Table 5.2. ZigBee is designed for low to very low duty cycles in static and dynamic environments with low data rate requirements and many active nodes; whereas Bluetooth is designed for high QoS, a variety of duty cycles with moderate data rate requirement in networks with limited active nodes.

Table 5.2   Comparison of application areas.

ZigBee	Bluetooth
Small packets over large networks.	Larger packets over small networks.
Mostly static networks with many, infrequently used devices.	Ad hoc networks with only a few devices.
Rapid network join time.	Long network join time.

Table 5.3 compares the air interface of the two standards, and Table 5.4 compares the timing characteristics. ZigBee devices can quickly attach, exchange information, detach, and go into sleep mode to achieve their objective of extending battery lifetime to up to two years. Bluetooth, which is targeted for devices such as PDAs and mobile phones that require regular charging, is incapable of such efficiency in power consumption.

5.6   Power Conservation Techniques

Because most sensor nodes rely on batteries, energy consumption must be managed efficiently. Sensor nodes must be in sleep mode approximately 99.9% of the time to ensure a very long battery life. Among the power management strategies taken are turning on the transceiver only during active communication, shutting down the central processing unit (CPU) between request processing and powering down the input and output (I/O) subsystems when not in use. A number of factors including application requirements and the hardware used determines the timing of the power down cycle.

The network stack reduces power consumption by eliminating communication or by turning off the radio when there are no communication needs. Possible approaches to save power include processing data locally and only communicating when an event of interest is detected. Performing aggregation can reduce communication and, hence, power consumption. For example, if an application has to determine the average temperature at shaded nodes in a certain geographic region, only a subset of nodes is selected for monitoring. The selection is performed at the

Table 5.3   Air interface comparison.

ZigBee	Bluetooth
DSSS	FHSS
11 chips/symbol	1,600 hops/s
62,500 symbols/s	1 million symbols/s
4 bits/symbol	1 bit/symbol

Table 5.4   Timing characteristics.

	ZigBee	Bluetooth
New slave enumeration	30 ms	3 s, typically 20 s
Sleep mode to AM	15 ms	3 s
Active slave channel access time	15 ms	2 ms

tree leaves, and nodes route their aggregated data upward. Each node transmits at most one packet to provide a statistical summary of its subtree.

Compression and scheduling conserve power at the lower levels. Control information can be piggybacked on data messages. Prescheduling reduces contention and the time the radio remains active. The network may assign specific responsibilities, such as retransmission, to certain nodes. The network may reject packets that are not of interest by turning off the radio after receiving only a portion.

The targeted sensor lifetime for many applications is between two and five years. A sensor node operating on one 1.5-V AA alkaline battery with an average power consumption in the range of 100 to 10 μW would last between two and seven years. A radio transceiver consumes tens of milliwatts. To conserve power, a node must be put in sleep mode most of the time; this can be achieved by using a duty cycle on the order of 0.1 to 1 percent while keeping a low sleep mode current no larger than the battery leakage current. Reducing power consumption requires optimization across all layers. The MAC layer plays the most crucial role in the overall communication protocols energy efficiency, especially for network with low duty cycling radios.

Microcontrollers running at 10 MHz operate at 1 mW. In standby mode, the power consumption is about 1 μW. If a device is active 1 percent of the time, its average power consumption is only a few microwatts. The source power can be obtained from a number of sources, for example, solar cells (10 to 100 μW/cm² indoors) and vibration of windows and air-conditioning ducts (100 μW). A typical operating point for WSNs is one year on a pair of AA batteries.

Power consumption can also be minimized by reducing and eliminating energy waste caused by:

- Idle listening: A node waits and listens for a packet to arrive.
- Overemitting: A node sends a message and the destination node is not ready to receive.
- Overhearing: A node listens for a message intended for another node.
- Collision: Two nodes transmit at the same time causing a collision. They have to retransmit the message later.
- Protocol overhead: Frame headers and signaling required by a MAC protocol.

Wireless Sensor MAC (WiseMAC) is a single-channel contention protocol based on nonpersistent CSMA (El-Hoiydi et al. 2004). It was designed to reduce power consumption for downlink transmissions in infrastructure WSNs. In this type of network, an AP has an unlimited power supply while the sensor nodes are constrained by battery power. A downlink transmission is a transmission from an AP to a sensor node. Because it is not possible to determine in advance when an AP would have a packet destined for a particular sensor node, the challenge is how to handle downlink transmission without requiring the sensor node to listen to the channel continuously as that would drain battery power. To accomplish this, we need to reduce or eliminate idle listening and overhearing.

Sensor nodes periodically sample the medium to check for transmission. This may be done to check if there is a packet intended for itself or to check if the medium is idle so that it may initiate transmission. All sensor nodes sample the medium periodically for a constant period, but their sampling schedule offsets are independent. If the medium is busy, the sensor node continues to listen until a data packet is received or until the medium becomes idle. Because there is no way for a node to ascertain if there is a packet pending, it has to continue listening until the medium is idle, consuming precious energy.

WiseMAC mitigates this problem by allowing the AP to learn the sampling schedule of all nodes so that it can schedule the transmission of a packet at the right time, thus reducing power consumption of the sensor node. The packet is preceded with a wake-up preamble to ensure that the node is awake when packet transmission starts. The AP maintains an up-to-date table of sampling schedules of all sensor nodes under its responsibility. A sensor node piggybacks information about its remaining time until its next scheduled sampling in acknowledgment packets to the AP. The AP uses this information to update the sampling schedule table. Collisions are not possible in the downlink channel because the AP is the only one that initiates transmission.

Polling is another mechanism that can be used to reduce power consumption. This is the approach used by the Periodic Terminal Initiated Polling (PTIP) protocol. Packets destined for sensor nodes are buffered at the AP, and sensor nodes regularly send POLL messages to retrieve the buffered packets. Upon receiving the POLL message, the AP returns the DATA packet, if there is one pending for the sensor node. Otherwise, it sends a short control packet. POLL messages are sent using CSMA. The interval between POLLs is determined randomly to avoid contention between synchronized nodes. When a sensor node receives a correct response to a POLL, it sleeps until the time for the next polling. It is not required to send an acknowledgment for the correct response. If the correct response is not received, it keeps sending POLLs until the correct response is received. The sequence number of the last correctly received DATA is piggybacked in the next POLL to inform the AP that the DATA packet has been received correctly and, hence, does not need to be retransmitted.

The IEEE 802.15.4 standard takes a similar approach to PTIP. The standard defines a PSM that reduces power consumption at the cost of higher delay. This scheme consists of a four-packet transmission: POLL-ACK-DATA-ACK. The AP buffers traffic destined for sensor nodes. A beacon that contains a traffic indication map (TIM) is transmitted periodically. TIM contains a list of sensor nodes with packets buffered at the AP. Sensor nodes wake up regularly to receive TIM, and if a node finds that it is on the list, it polls the AP to retrieve the buffered packet. When the AP receives the POLL message, it returns an ACK, which instructs the sensor node to remain in listen mode. The AP then finds the correct packet and sends the DATA packet to the sensor node. Upon receiving the packet, the sensor node returns an ACK.

Both WiseMAC and PSM makes use of a more bit in the data packet header to indicate that there are one or more data packets queued in the buffer for the sensor node. If the more bit is set to 1, it indicates that the sensor node should remain in listen mode after sending an acknowledgment as the next packet will be sent. PSM requires the sensor node to send another POLL to retrieve the next packet.

The various power conservation approaches may conflict with one another. A combination of different approaches may be deployed based on the requirements of an application.

5.7   Network and Communications

Unlike traditional Internet applications, sensor network applications require protocols that are optimized for their unique communication patterns. Standards for sensor networks are 802.15.4 and the ZigBee protocol. The 802.15.4 standard specifies the RF channel and signaling protocol. ZigBee is built on top of 802.15.4 and specifies the application-level communication protocol between devices. The 802.15.4 standard determines which radio hardware to use, and ZigBee determines the content of the messages transmitted.

Unlike communications on the Internet, communications in WSNs are usually performed in aggregate. Participants are identified by attributes such as physical location or sensor value range. This style of routing, known as directed diffusion, is a process where nodes express interest in data by attribute. WSNs deploy disruption-tolerant networking (DTN), where bundles of data are transferred reliably hop by hop. The DTN model better suits the variable connectivity resulting from dynamic environments and the need to duty cycle.

Each sensor node has a radio that provides communication links to neighboring nodes. They perform a distributed algorithm to determine how to route data. The physical placement determines connectivity but other variables, such as obstruction, interference, and antenna orientation, make it difficult to determine connectivity a priori.

Sensors use radio broadcasts. The lowest layer controls the physical radio device. When one node transmits, the signal is received by a set of neighboring nodes unless it is garbled by another transmission. The link layer is responsible for avoiding contention for the radio channel. It listens on the channel and transmits only when the channel is clear, which is similar to the way CSMA/CD works. When a node is not transmitting, it scans the channel for transmissions from other nodes. The packet layer manages buffers, schedules packets onto the radio, detects or corrects errors, handles packet losses, and forwards packets to system or application components.

Disseminating information involves a flooding protocol. A root node broadcasts a packet containing an ID. Receiving nodes retransmit the packet so that distant nodes can receive it. A node may receive different versions of the same message from different neighboring nodes. The ID in the packet is used to detect and discard duplicate packets. The network uses dissemination to issue commands, send alarms, and configure the network. Dissemination is also used to establish routes. Each packet identifies the transmitter and its distance from the root. A distributed tree is formed by identifying a node closer to the root. The network uses the reverse communication tree for data collection by routing data back to the root or for data aggregation by processing data at each level of the tree. The root may act as a gateway to a more powerful network or as an aggregation point within the sensor network.

For data from the sensors to be sent over long distances, data have to be routed hop by hop through nodes. Because communication is one of the most energy consuming operations, with each bit consuming as much energy as 1000 instructions, WSNs should process data within the network whenever possible.

5.8   Configuration of Sensor Networks

Adaptive Self-Configuring sEnsor Networks Topologies (ASCENT) is a mechanism that allows an adaptive self-configuration of sensor networks (Cerpa and Estrin 2004). As the density of sensors in an area increases, it is unnecessary for all nodes to be active simultaneously. Allowing the nodes to self-configure themselves by determining which nodes can be active while the others sleep allows the nodes to conserve energy that is critical for sensor nodes. ASCENT takes an adaptive approach because a centralized solution would pose all the restrictions of a centralized system, especially in terms of scalability and robustness. Because a single, central node cannot sense the conditions of other nodes, it has to be kept updated of the status of other nodes, thus, generating much traffic and consuming more power to transmit or receive the updates. This mechanism assumes the following conditions:

■ Ad hoc deployment: It is unlikely that sensor nodes are uniformly distributed. Even if they are, it does not necessarily mean there is uniform connectivity due to unpredictable propagation effects.

■ Energy constraints: Sensor nodes operate on battery power. It is imperative that energy consumption be minimized to prolong their lifetimes.
■ Unattended operation under dynamics: The high number of nodes in a system rules out manual configuration. Environmental dynamics precludes design-time preconfiguration.

ASCENT adaptively selects active nodes from all nodes in the network. An active node stays awake and performs multihop packet routing. The remaining nodes remain passive and check periodically if they should become active. Referring to Figure 5.10a, initially, only a few nodes are active while the others are passively listening to packets but not transmitting. The source starts transmitting data packets toward the sink, but because the sink is at the limit of the radio range, there is very high packet loss from the source. This is termed a communication hole. To overcome this problem, the sink sends help messages to signal neighbors that are in listen-only mode, that is, passive neighbors, to join the network.

When a neighbor receives a help message, it may decide to join the network (Figure 5.10b). When a node joins the network, it starts transmitting and receiving packets, that is, it becomes an active neighbor. As soon as a node joins the network, it announces its active status to other passive neighbors by sending a neighbor announcement message. This continues until the number of active nodes stabilizes at a certain value and the cycle stops (Figure 5.10c). When the process is completed, the group of newly active neighbors that joined the network results in a more reliable data delivery from source to sink. The process restarts when a future network event (e.g., node failure) or environmental effect (e.g., new obstacle) causes an increase in packet loss.

A node may be in a sleep, passive, test, or active state. When a node is initially deployed, a random timer is turned on to avoid synchronization, and it is initialized in a test state. Nodes in the test state exchange data and routing control messages. A node in test state sets a timer T_t and sends neighbor announcement messages. If before T_t expires, the number of active neighbors exceeds the neighbor threshold (NT) or if the average data loss (downlink) rate is higher than the average loss before entering in the test state, the node moves into passive state. Otherwise, the node enters the active state when T_t expires. The purpose of the test state is to determine if the addition of a new node would improve connectivity.

When a node enters the passive state, it sets up a timer T_p and sends new passive node announcement messages that are used by active nodes to estimate the total density of nodes in the neighborhood. Active nodes transmit the density estimate to new passive nodes in the neighborhood. When T_p expires, the node enters the sleep state. If the following events occur before T_p expires, the node moves into the test state:

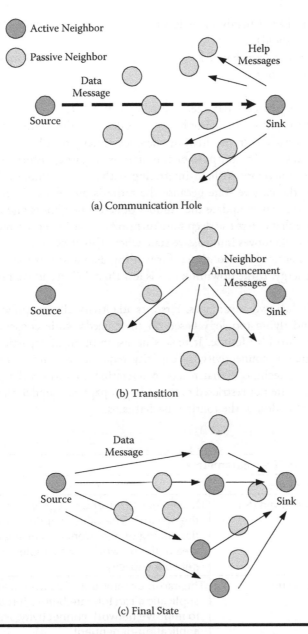

(a) Communication Hole

(b) Transition

(c) Final State

Figure 5.10 Network self-configuration (From Cerpa, A., and Estrin, D., 2004. ASCENT: Adaptive Self-Configuring sEnsor Networks Topologies. *IEEE Transactions on Mobile Computing* 3(3):272.) Used with permission.

■ The number of neighbors is less than *NT* and *downlink* is greater than *LT* (loss threshold).

■ *Downlink* is less than *LT,* but the node received a help message from an active neighbor.

Table 5.5 summarizes the parameters used by ASCENT.

When in the passive (listen-only) state, nodes have their radios on and are able to overhear all packets transmitted by their active neighbors but they do not route or forward packets. The purpose of this state is to gather information regarding the state of the network without interfering with the other nodes. Energy is still consumed in the passive state because the radio is on. Nodes in the passive and test states continuously update the number of active neighbors and downlink rate values. A node that enters the sleep state turns off the radio, sets a timer T_s and goes to sleep. The node moves into passive state when T_s expires.

A node in active state continues forwarding data and routing packets until it runs out of energy. If the downlink rate is less than *LT,* the active node sends help messages.

ASCENT has two advantages. First, its adaptivity allows applications to configure the underlying topology based on their needs while conserving power to extend the network's lifetime. It makes no assumption of a particular model of fairness, degree of connectivity, or capacity required. Second, it makes use of a self-configuring technique that reacts to operating conditions that are measured locally and that are not restricted to the radio propagation model, the geographical distribution of nodes, or the routing mechanisms.

Table 5.5 ASCENT parameters.

Parameter	Description
Neighbor threshold (NT)	Determines the average degree of connectivity of the network. The value is updated dynamically depending on the events occurring in a certain area of the network, for example, to increase network capacity.
Loss threshold (LT)	The maximum amount of downlink an application can tolerate before it requests help to improve network connectivity. The value is application-dependent.
Test timer (T_t)	The maximum time a node remains in test state.
Passive timer (Tp)	The maximum time a node remains in passive state.
Sleep timer (T_s)	The amount of time a node sleeps to conserve energy.

5.9 WSN and Emergency Response Applications

Section 5.1 gives an overview of WSN applications in environment monitoring. Another important application is in emergency services and response. In a disaster scenario, quick and efficient action by first responders is critical to saving lives. An important aspect in disaster scenarios is that we cannot assume that any communication infrastructure will be available. For example, after an earthquake, there is a good chance that the existing communication infrastructure would be destroyed. Consequently, a wireless communication infrastructure that can be deployed quickly and with the least hassle is imperative to the search and rescue operation.

The emergency response applications discussed below are still in the experimental stage, but the discussion gives you an idea of the great potential and the important role WSNs will play in the near future.

5.9.1 CodeBlue Project

CodeBlue is a project at Harvard University that provides an infrastructure to support an emergency response application in disaster scenarios such as an earthquake (Lorincz et al. 2004). It is an integration of sensor nodes and various types of mobile devices to provide support for search and rescue operations. The sensor motes used for this application are different from a mobile device such as a PDA that has higher CPU speed, memory size, and communication capabilities. The motes have limited bandwidth and computational power, which rules out the use of common IPs. Support for the application comprises providing the facilities to form ad hoc networks, resource naming and discovery, security, and in-network aggregation of sensor-produced data.

The setting up of an ad hoc network is important to allowing the mobile devices to route data among themselves and from the disaster venue to a control center that then disseminates the relevant information to the emergency personnel. Because the devices have to share the limited bandwidth offered by 802.15.4 and ZigBee, data has to be prioritized. Critical data, such as data from a vital signs monitor from a victim in critical condition, is given priority.

Tracking rescue personnel and patients is an important aspect of the application. For example, firefighters wear tags to allow easy tracking of their movements to coordinate search and rescue operations more effectively. The firefighters can be informed if a particular section of a building is unstable and about to collapse and directed to evacuate immediately. The RF-based tracking system is called MoteTrack and operates using low-power, single-chip transceiver. The beacon nodes for use in conjunction with MoteTrack can be used to replace existing smoke detectors so that it has a two-in-one function of detecting smoke and tracking location. The sensor nodes broadcast periodic messages consisting of a tuple {sourceID, powerLevel}, where sourceID is the node unique identifier and powerLevel is the transmission power level used to transmit the message.

In cases where there is a high number of casualties, tractable vital signs sensors can be attached to the patients to aid quick location of a patient who suddenly needs immediate attention. Two types of vital signs sensors have been developed: a pulse oximeter and an ECG monitor. A pulse oximeter observes a patient's heart rate and blood oxygen saturation by measuring the amount of light transmitted through a noninvasive sensor attached to his finger. The EKG monitor observes the heart's electrical activity in patients who are severely injured. It is used to detect arrhythmia (abnormal heart beat rhythm) or ischemia (lack of blood flow and oxygen to the heart) that indicates a potentially serious condition. Information from these sensor motes are transmitted to a PDA carried by first responders. The PDA is able to display vital signs information from more than one patient.

A patient's medical record is confidential and its privacy must be guarded at all times. Devices and data must still be protected against common attacks, such as spoofing and denial of service (DoS) attacks; therefore, the handling of security and privacy issues for an emergency response application involves a new set of issues and conditions that do not exist in traditional applications. A search and rescue operation often involves personnel from different organizations, and it is unreasonable to expect that they have exchanged security or configuration information a priori. It is also unreasonable to expect first responders to spend time typing passwords to set access permission to a database or to require them to do any manual configuration as it would distract them from critical tasks. In a life-threatening situation, denying or delaying a legitimate user access to data due to authentication procedures is not an option.

Consequently, CodeBlue makes use of a best-effort security model and its architecture provides support for an ad hoc network security model that does not require manual configuration and has the capability to self-organize participating devices. The system is designed to automatically cope with nodes joining and leaving the network. The security model provides seamless credential handoff from one first responder to another to grant access to the other, without requiring a preexisting relationship between them. This would be required when a patient is brought from the emergency site to the hospital, where access rights to the patient's data must be transferred from the first responder to the medical team waiting at the hospital.

Data encryption is performed using Elliptic Curve Cryptography (ECC). ECC is chosen over RSA because it uses smaller keys and is more computationally efficient. The implementation of ECC on Mica2 generates a key in 35 s. Even though it is not negligible, it is acceptable if a key is not generated frequently. Mica2 is a sensor mote that is based on a 7.3-MHz Atmel Atmega128L embedded controller with 4 Kbytes of RAM and 128 Kbytes of read-only memory.

A mobile node carried by or attached to a first responder who wishes to determine its location has to listen for a certain period to acquire a signature. A signature is a combination of the beacon messages received over an interval and the received signal strength indication (RSSI) for each transmission power level. The mobile node acquires a reference signature set at known, fixed locations throughout the building. It then sends the signature to the beacon node from which it receives the strongest RSSI to estimate its location.

MoteTrack uses a decentralized approach to location tracking. It replicates its reference signatures such that a beacon node stores only a subset of all reference signatures. A beacon node stores reference signatures and performs computations based on the locally stored data. This decentralized approach is prone to failure and is further enhanced by the use of an adaptive algorithm to estimate location.

CodeBlue was designed specifically to assist first responders in the event of natural disasters. However, the following application, which utilizes sensors and robots, has a wider scope and can also be used in accident scenarios, such as monitoring a chemical spill.

5.9.2 WSN and Robots in Disaster Response Applications

In disaster scenarios, it might not always be possible to send humans to certain places, for example, because the area is too small or too dangerous. An approach to address this problem deploys a combination of small mobile robots and sensor nodes. To examine the effectiveness of combining robots and sensors to assist firemen, a group of researchers deployed a network of stationary mote sensors, mobile robots equipped with cameras, and stationary radio tags to assist firefighters in their tasks (Kumar et al. 2004). Communications between robots make use of the 802.11b network.

The robots and sensor nodes, termed agents, may be static, autonomous, or tele-operated by humans. During a search and rescue operation, first responders and tens of agents enter a building, most probably with little, if any, knowledge of the building. If floor plans are available, agents make use of them to expedite the search process. The agents autonomously organize themselves to communicate effectively, integrate information efficiently, and obtain relative position information. Among data gathered are temperatures, toxin concentrations, gas leaks, sources of danger, and locations of victims. The utilization of robots to assess the level of danger avoids unnecessary risk to the life of the rescue team—it is more acceptable to lose a robot than a human life. An agent is equipped with a microphone so that a trapped victim can call for help. The location of the victim is identified and the task of rescuing the victim may commence. If an area is perceived as unsafe, for example, temperature has exceeded 100°C, it is marked and firefighters are warned not to go into this area.

In a scenario where an accident involves a lorry carrying hazardous material, a similar network can be deployed to determine the extent of contamination in the area. The robots are used to follow and monitor the spread of the chemical spill.

The network of sensors and robots comprises low-cost radio tags with conventional line-of-sight optical sensors that operate in the IR spectrum. Each tag is assigned a unique ID. A robot identifies tags in the vicinity by sending a query message. Any robot that receives the query sends a reply. The robot estimates its distance to the others based on the time elapsed between sending the query and receiving a reply. The sensors store information about the area they cover locally. The information may also be sent to a BS for analysis and later used in decision making. Localized information about the relative location of agents is used to build a global map of sensory information that is used to navigate humans and robots to a target while avoiding dangerous areas. This navigation guide application is composed of simple nodes distributed over a large geographical area and is used to assist in accomplishing global tasks.

Because the system consists of numerous agents and it must be robust enough to cope with the addition and disappearance of agents, each agent is anonymous. Each agent must operate asynchronously and makes no assumption of who its neighbors are and the location of neighbors.

An important requirement of robot and sensor systems in emergency application is that they should be autonomous and require minimal human supervision and configuration to avoid distracting emergency personnel from more important tasks.

5.10   Summary

Even though sensors initially found widespread use as tools to monitor the environment, their applications have now penetrated other areas, such as search and rescue, smart homes, and sports. Remote monitoring and management are important aspects of WSNs not only because it might be impossible to visit isolated locations regularly, but also because remote management is often an intrinsic characteristic of the application itself: In a smart home application, homeowners may wish to monitor their homes while away on vacation. As people who use a WSN may not always be people with the technical background to install complex devices with complex interfaces, environmental sensor networks must primarily consist of off-the-shelf components that are easy to deploy, maintain, and understand. Sensor nodes must also be low-cost, unobtrusive, durable, energy efficient, and maintenance-free.

References

Cerpa, A., and D. Estrin. 2004. ASCENT: Adaptive Self-Configuring sEnsor Networks Topologies. *IEEE Transactions on Mobile Computing* 3(3):272.

Culler, D., D. Estrin, and M. Srivastava. 2004. Overview of sensor networks. *Computer* 37(8):41.

El-Hoiydi, A., and J. D. Decotignie. 2004. Low power MAC protocols for infrastructure wireless sensor networks. *Mobile Networks and Applications* 10(5):675–690.

Kumar, V., D. Rus, and S. Singh. 2004. Robot and sensor networks for first responders. *IEEE Pervasive Computing* 3(4):24.

Lorincz, K., D. J. Malan, T. R. F. Fulford-Jones, A. Nawoj, A. Clavel, V. Shnayder, G. Mainland, M. Welsh, and S. Moulton, Sensor networks for emergency response: challenges and opportunities. *IEEE Pervasive Computing* 3(4):16.
Martinez, K., J. K. Hart, and R. Ong. 2004. Environmental sensor networks. *Computer* 37(8):50.

Bibliography

Callaway, E., P. Gorday, L. Hester, J. A. Gutierrez, M. Naeve, B. Heile, and V. Bahl. 2002. Home networking with IEEE 802.15.4: A developing standard for low-rate wireless personal area networks. *IEEE Communications Magazine* 40(8):70.
Das, S. K., D. J. Cook, A. Bhattacharya, E. O. Heierman, III, and T. Y. Lin. 2002. The role of prediction algorithms in the MAVHome Smart Home Architecture. *IEEE Wireless Communications* 9(6):77.
Enz, C. C., A. El-Hoiydi, J. D. Decotignie, and V. Peiris. 2004. WiseNET: An ultralow-power wireless sensor network solution. *Computer* 37(8):62.
Evans-Pughe, C. 2003. Is the ZigBee wireless standard, promoted by an alliance of 25 firms, a big threat to Bluetooth? *IEE Review* 49(3):28.
Gutierrez, J. A, M. Naeve, E. Callaway, M. Bourgeois, V. Mitter, and B. Heile. 2001. IEEE 802.15.4: A developing standard for low-power low-cost wireless personal area networks. *IEEE Networks* 15(5):12.
Hill, J., M. Horton, R. Kling, and L. Krishnamurthy. 2004. The platforms enabling wireless sensor networks. *Communications of the ACM* 47(6):41.
Poole, I. 2004. What exactly is ZigBee? *IEE Communications Engineer* 2(4):44.
Szewczyk, R., E. Osterweil, J. Polastre, M. Hamilton, A. Mainwaring, and D. Estrin. 2004. Habitat monitoring with sensor networks. *Communications of the ACM* 47(6):34.
Waldo, J. 2005. Embedded computing and Formula One racing. *IEEE Pervasive Computing* 4(3):18.
Zheng, J., and M. J. Lee. 2004. Will IEEE 802.15.4 make ubiquitous networking a reality? A discussion on a potential low power, low bit rate standard. *IEEE Communications Magazine* 42(6):140.

Online Resources

CodeBlue Project. http://www.eecs.harvard.edu/~mdw/proj/codeblue/ (Accessed February 8, 2007).
IEEE 802.15 WPAN Task Group 4 (TG4). http://www.ieee802.org/15/pub/TG4.html (Accessed February 8, 2007).
ZigBee Alliance. http://www.zigbee.org/en/index.asp (Accessed February 8, 2007).

Chapter 6

Mobile Ad Hoc Networks

A mobile ad hoc network (MANET) is formed when two or more devices form a peer-to-peer network with no centralized control. This may be used in meetings. A MANET can be used in the field by search and rescue or military personnel. A MANET allows the exchange of data between people working in a team. Because there is no centralized control, nodes in an ad hoc network cooperate to deliver packets to one another. To deliver a packet to a destination node, an originating node has to discover a route to the destination using a routing protocol. The infrastructureless, decentralized nature of MANETs makes an efficient routing protocol critical to facilitating communications between nodes.

Routing protocols can be classified into two categories:

1. Reactive protocols: A mobile node initiates route discovery only when it needs to deliver a packet to a destination. There are two reactive protocols: ad hoc on-demand distance vector (AODV) and dynamic source routing (DSR).
2. Proactive protocols: Nodes continually exchange routing information so that it is readily available when a node would like to send packets to another node. There are two proactive protocols: optimized link state routing (OLSR) and topology broadcast based on reverse path forwarding (TBRPF).

6.1 AODV

AODV (Perkins and Das 2003) provides a dynamic, self-starting, multihop routing between mobile nodes to establish and maintain an ad hoc network. Routes are only maintained for destinations in active communication. The algorithm provides a way for mobile nodes to respond to link breaks and changes in network topology in a timely manner. The algorithm also establishes a route between a source and a destination that is loop-free. It is designed to scale for thousands of mobile nodes and can handle low, moderate, and relatively high mobility rates with a variety of traffic levels.

A destination SN is associated with each route entry in the routing table. The SN is generated by the destination and is included with other route information sent to requesting nodes. When there is more than one route available to a destination node, the one with the greatest SN is chosen.

6.1.1 Message Formats

AODV defines three types of messages, namely route request (RREQ), route reply (RREP), and route error (RRER). The messages are sent to port 654 using User Datagram Protocol (UDP). The message format of RREQ is shown in Figure 6.1. The fields in the message are:

- Type: For RREQ, the value is set to 1.
- Flag: The five flag bits are J (join), R (repair), G (gratuitous), D (destination only), and U (unknown). The J and R flags are reserved for multicasts. The G flag indicates whether a gratuitous RREP should be unicast to the node specified in the destination IP address field. When the D flag is set to 1, it indicates that only the destination may respond to the RREQ. When the U flag is set to 1, it indicates that the destination SN is unknown.
- Reserved: Set to 0 and is ignored on reception.
- Hop count: The number of hops from an originator to the node handling the request.
- RREQ ID: Taken in conjunction with the originator's address, it uniquely identifies a RREQ.

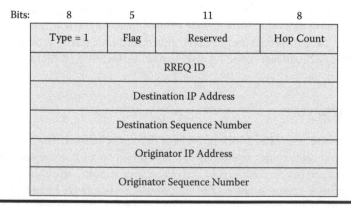

Bits: 8	5	11	8
Type = 1	Flag	Reserved	Hop Count
RREQ ID			
Destination IP Address			
Destination Sequence Number			
Originator IP Address			
Originator Sequence Number			

Figure 6.1   RREQ message format.

Destination IP address: The address of the destination.

Destination SN: The latest SN previously received by the originator for any route toward the destination.

Originator IP address: Address of the originator that sent the RREQ.

Originator SN: The current SN to be used in the route entry pointing toward the originator of the route request.

The message format of RREP is shown in Figure 6.2. The fields in the message are:

Type: Set to 2.

R: Repair flag used for multicasts.

A: When set to 1, this indicates that an acknowledgment is required.

Reserved: Set to 0.

Prefix size: A nonzero value means the indicated next hop can be used for any node with the same routing prefix as the requested information.

Hop count: The number of hops from the originator to the destination.

Destination IP address: The address of the destination.

Destination SN: The SN associated to the route.

Originator IP address: Address of the originator that sent the RREQ.

Lifetime: Time in milliseconds for which the nodes receiving the RREP consider the route to be valid.

The message format of RRER is shown in Figure 6.3. The fields in the message are:

Type: Set to 3.

N: No delete flag. When the value is set to 1, it means a node has performed a local repair of a link and an upstream node should not delete the route.

Reserved: Set to 0.

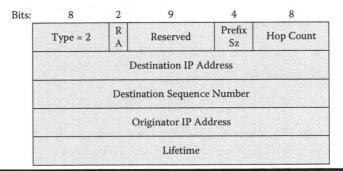

Figure 6.2   RREP message format.

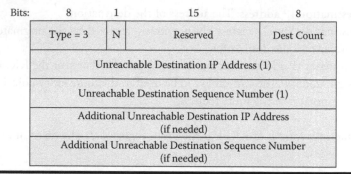

Figure 6.3 RRER message format.

■ DestCount: The number of unreachable destinations in the message. The value must be at least 1.

■ Unreachable destination IP address: The IP address of the destination that has become unreachable due to a link break.

■ Unreachable destination SN: The SN in the route table entry for the destination listed in the previous unreachable destination IP address field.

6.1.2 Destination SN

The destination SN plays an important role in ensuring that a route is loop-free. It is updated each time a node receives new information about the SN from RREP, RREQ, or RRER messages related to a destination. Immediately before a node initiates a route discovery, it increments its own SN to avoid conflicts with previously established reverse routes toward the originator of a RREQ. Immediately before a destination node returns a RREP, it updates its own SN to the maximum of its current SN and the destination SN in the RREQ packet. A node determines that information about a destination is not stale by comparing its current stored SN with that obtained from an incoming AODV message. If the value of the SN in the message is greater than its stored value, the information related to the destination is considered stale and discarded.

6.1.3 Routing Table

An entry in a routing table contains the following information:

■ Destination IP address: Address of the destination node
■ Destination SN: The latest SN for the destination node
■ Valid destination SN flag
■ Other state and routing flags

- Network interface
- Hop count: The last known hop count to the destination
- Next hop: The node to forward a data packet to in order to reach the destination
- List of precursors
- Lifetime: How long a route remains valid

When a node receives an AODV message from a neighbor, it checks its routing table for an entry of the destination. If an entry for that destination does not exist, one is created. If an entry exists, it compares the destination SN in the entry and in the message. The route is updated if:

- The SN in the table entry is lower than the one in the message.
- The SNs are equal, but the hop count plus one is smaller than the existing hop count in the table.
- The SN is unknown.

The lifetime field in the routing table is determined from the AODV message or is initialized to ACTIVE_ROUTE_TIMEOUT. The lifetime of an active route is updated each time the route is used. Each time a route is used to forward a data packet, each node on the route (including the originator and destination) updates this field to the current time plus ACTIVE_ROUTE_TIMEOUT.

6.1.4 Route Discovery

Referring to Figure 6.4, if node 1 (originator) wants to deliver a data packet to node 2 (destination), and it does not know the route to node 2, it initiates a route discovery by broadcasting a RREQ. Node 1 may also initiate a route discovery if its routing table indicates that the route to node 2 has been marked as invalid.

The destination SN in RREQ is copied from the entry in the routing table. If the SN is unknown (e.g., an entry for node 2 does not exist in the routing table), the U flag is set to 1. Node 1 increments its originator SN and inserts it in the RREQ. The hop count field is set to zero. Before node 1 broadcasts the RREQ message, it buffers the RREQ ID and originator IP address for a period of PATH_DISCOVERY_TIME. When node 1 receives a RREQ from a neighbor, it compares the RREQ ID and originator IP address value in the message to the ones in its buffer. If it recognizes it as the RREQ it has generated, it does not process nor forward the RREQ.

A route is determined when the RREQ reaches node 2. When node 2 receives the RREQ, it returns a RREP to node 1. A reverse route is set up to forward a RREP back to the originator. Alternatively, if an intermediate node, let us say node 7, has recently communicated with node 2 and it knows how node 2 can be

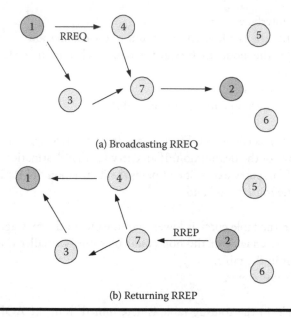

(a) Broadcasting RREQ

(b) Returning RREP

Figure 6.4   Route discovery.

reached, it returns RREP to node 1 and does not forward the RREQ. Node 7 can only return a RREP if it has a fresh-enough route—a valid route entry for the destination whose SN is at least equal to that contained in the RREQ. Intermediate nodes (e.g., node 3 and node 4) that receive the RREP cache a route back to the originator of the request. The RREP is sent by unicast. Node 1 stores the route to node 2 in its routing table. A route to a destination is said to be an active route if the route to the destination is marked as valid. Only active routes can be used to forward data packets.

If node 1 does not receive a RREP after NET_TRAVERSAL_TIME milliseconds, it may broadcast another RREQ. The maximum number of retries is determined by RREQ_RETRIES and the number of RREQs generated by node 1 should not exceed RREQ_RATELIMIT per second. The value of RREQ ID is increased for each RREQ generated. If after RREQ_RETRIES, node 1 still does not receive a RREP, data packets destined for node 2 are dropped and a destination unreachable message is delivered to the application. To reduce congestion, node 1 must execute a binary exponential backoff before attempting to broadcast another RREQ for node 2.

When a node receives a RREQ, it checks if it has received a RREQ with the same originator IP address and RREQ ID within the last PATH_DISCOVERY_TIME. If it has, the RREQ is discarded. Otherwise, it searches for a reverse route to the originator. If an entry for the originator does not exist in its routing table, one is created. Otherwise, the entry is updated using the originator SN from the

RREQ. The reverse route is needed if the node receives a RREP back to the node. The following four steps are taken when creating or updating the reverse route:

1. If the originator SN in the RREQ is greater than the one in the table entry, it is copied into the entry.
2. The valid SN field is set to true.
3. The next hop is set to the node from which the RREQ is received.
4. The HOP COUNT in RREQ is copied into the entry.

If the node does not know the route to the destination (therefore, it cannot generate a RREP) and the time-to-live (TTL) value is greater than 1, it updates the RREQ by increasing the hop count by one and decreasing the TTL by one. Then it broadcasts the RREQ on all of its configured interfaces.

A node generates a RREP if either it is the destination or it is an intermediate node with an active route to the destination. When a node generates a RREP, it discards the RREQ. If an intermediate node generates a RREP, the destination might not receive a RREQ, and therefore, is unable to learn the route to the originator. To ensure that the destination learns the route to the originator, the originator should set the G flag in the RREQ. This requires the intermediate node that generates the RREP to unicast a gratuitous RREP to the destination, thus creating a bidirectional communication between the originator and the destination.

The node generating the RREP copies the destination IP address and the originator SN from the RREQ into the corresponding field in RREP. If the destination generates the RREP, it increments its SN by one if the SN in the RREQ is equal to that incremented value. The SN is copied into the destination SN field in the RREP and the hop count field is set to 0. It then copies its MY_ROUTE_TIMEOUT value into the lifetime field. If an intermediate node generates the RREP, it copies its known SN for the destination into the destination SN in RREP.

When a node receives a RREP message, it searches for a route to the previous hop. Then it increases the hop count in RREP by one. The forward route for the destination is created if one does not already exist. The node compares the destination SN in the message with one in its routing table entry. The entry is updated if the SN is marked as invalid. It is also updated if:

- The destination SN in the RREP is greater than the one in the entry and the known value is valid.
- The SNs are the same, but the route is marked as inactive.
- The SNs are the same and the new hop count is smaller than the one in the entry.

If the entry is created or updated, the following six steps are taken:

1. The route is marked as active.
2. The destination SN is marked as valid.

3. The next hop is assigned to be the node from which the RREP is received.
4. The hop count is set to the value of the new hop count.
5. The expiry time is set to the current time plus the value of lifetime in RREP.
6. The destination SN is set to the value of the destination SN in RREP.

The node can now use this route to forward data packets to the destination.

6.1.5 Maintenance of Active Routes

Nodes monitor the link status of next hops in active routes. If a node detects that a link in an active route is broken, it sends a RRER message to other nodes to notify them that the link is lost. The RRER message indicates destinations that are no longer reachable due to the broken link. To facilitate this reporting mechanism, each node maintains a precursor list that contains the addresses of neighbors that are likely to use it as a next hop toward a destination. When a destination is unreachable, the corresponding entry in the routing table is marked invalid.

6.2 DSR

DSR (Johnson et al. 2004) is a reactive routing protocol that is designed specifically for use in multihop wireless ad hoc networks of up to 200 nodes. It is designed for very high rate mobility. It incurs very low overhead but reacts very quickly to changes in the network. DSR allows an ad hoc network to be self-organizing and self-configuring without relying on an existing network infrastructure or administration. Mobile nodes cooperate to forward packets to each other to allow communications over multiple hops between nodes that are not within transmission range of each other. The routing information is automatically updated as nodes move about, join, or leave the network. DSR is composed of two mechanisms that operate on demand:

1. Routing discovery: A mechanism that allows a node S that wishes to send a packet to a destination D to obtain a source route to D. Route discovery is invoked only when S attempts to send a packet to D and it does not know the route to D.
2. Route maintenance: While node S is using a source route to node D, S is able to detect if the network topology has changed such that it can no longer use its route to D using route maintenance. If route maintenance indicates that the source route is broken, S can attempt to use an alternative route to D, or it can invoke route discovery to find a new route to D. Route maintenance is only used when S is sending packets to D.

DSR incurs low overhead because it does not require periodic exchange of packets to maintain routing information. If nodes remain relatively stationary, the overhead packets drops to zero. As the rate of movement increases, the routing overhead scales to only what is needed to track the routes currently in use. The protocol does not react to network topology changes that do not affect currently used routes.

The protocol assumes that each node in the ad hoc network is willing to forward packets to other nodes in the network. The diameter of an ad hoc network is defined as the minimum number of hops necessary for a packet to reach any node located at one extreme edge of the network to another node located at the opposite extreme. The diameter is assumed to be small (between 5–10 hops).

6.2.1 Route Discovery

DSR uses explicit source routing where the packet header contains the complete, ordered list of nodes through which the packets will pass to reach its destination. This approach allows a source node to select and control the routes for its packets, supports the use of multiple routes to a destination, and guarantees that routes are loop-free. This approach also allows other nodes forwarding or overhearing the packet to cache the routing information for future use.

The sender obtains a suitable route to the destination by searching its route cache of routes previously learned. If none is found, it initiates a Route Discovery protocol to find a new route to the destination. The node that initiates the Route Discovery protocol is called the initiator and the destination node is the target.

Suppose node A is trying to discover a route to node E. To initiate route discovery, A broadcasts a RRREQ that is received by all nodes within the transmission range of A. Each RREQ contains information about the initiator and the target and a unique request ID (ID=2 in Figure 6.5). Each RREQ contains a record listing the address of each intermediate node to which the request is forwarded. Initially, the route record contains only node A (the initiator). Figure 6.5 shows the content of the route record.

When a node receives a RREQ, it checks if it is the target. If it is not the target, it checks:

■ If it has recently received a RREQ with the same ID and target from A.
■ If its own address is in the route record.

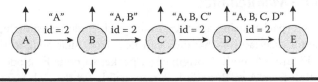

Figure 6.5   Route discovery process.

If either is true, the node discards the RREQ. Otherwise, the node appends its own address to the route record and forwards the request as a local broadcast packet. The node updates its own cache using information in the route record, if necessary.

When node E (the target) receives the RREQ and identifies itself as the target, E returns a RREP to the initiator of the route discovery (node A). The reply contains a copy of the accumulated route record from RREQ. When A receives the RREP, it caches the route information in its route cache for use to send packets to E in the future.

When E returns a RREP, it also examines if its own route cache contains a route for A. If one is found, it uses it as the source route to deliver the RREP. Otherwise, E has to perform its own route discovery for node A. In this case, the RREP is piggybacked in the packet that contains its RREQ for A. Alternatively, E could simply reverse the sequence of hops in the route record that it is copying into the RREP and use it as the source route for the RREP.

An intermediate node that is not the target node may return a RREP. Suppose node C searches its route cache for a route to the target when it receives the RREQ and finds one. C returns a RREP to A and does not forward the RREQ. The source route returned in the reply is a concatenation of the route record in the RREQ and the source route obtained from C's route cache. However, C must ensure that the concatenated route record returned in the RREP does not contain duplicate nodes.

When A initiates the route discovery, it saves a copy of the packet that it could not send to E because it does not have routing information in a send buffer. A packet in the send buffer is discarded after a predefined period expires to prevent the buffer from overflowing. While the packet is in the send buffer, A occasionally initiates a new route discovery for the packet's destination address. To reduce the overhead of route discoveries, a node executes an exponential backoff algorithm to limit the rate that it initiates a new route discovery for the same target. The timeout between successive discoveries is doubled.

A RREQ contains a hop limit that is implemented using the TTL field in the IP header of the packet that carries the RREQ. The hop limit is used to limit the number of intermediate nodes allowed to forward the RREQ. Each node that receives the RREQ decrements the hop limit by one and the packet is discarded when the limit reaches 0.

6.2.2 Route Maintenance

When sending a packet using a source route, each node transmitting the packet is responsible for ascertaining that data can flow from that node to the next hop. Referring to Figure 6.6, node A originates a packet to node E. Node A is responsible for the link from A to B, node B is responsible for the link from B to C and so on. An acknowledgment is returned to confirm that a link is capable of carrying

Figure 6.6   Packet forwarding.

data. If a built-in acknowledgment mechanism is not available, a node can explicitly request the next hop node to return a DSR-specific software acknowledgment.

After receiving an acknowledgment, a node may choose to not require acknowledgments from that neighbor for a brief period. In case a software acknowledgment is requested, the request should be retransmitted up to a maximum number of times. Retransmissions of the acknowledgment request can be sent as a separate packet, piggybacked on a retransmission of the original data packet, or piggybacked on any packet with the same next-hop destination. If no acknowledgment is received after the request has been transmitted the maximum number of times, the sender assumes the link to the next hop node is broken. The node removes this link from its route cache and returns a RRER to all nodes that have sent a packet via that link since the last acknowledgment was received.

6.3   OLSR

OLSR (Clausen and Jacquet 2003) is a proactive, decentralized, link-state routing protocol. It is a hop-by-hop routing, meaning that each node uses its local information to route packets. It is suitable for large and dense mobile networks, and networks where the traffic is random and sporadic between a larger set of nodes. OLSR is independent from other protocols and makes no assumption about the underlying link layer.

OLSR does not require reliable transmission of control messages. Because nodes transmit messages periodically, OLSR can sustain a reasonable loss of some messages. Each control message contains a SN to allow identification in case messages have been reordered during transmission. OLSR provides protocol extensions, such as sleep mode operation and multicast routing, that allow additions to the protocol without affecting backward compatibility.

Each node selects a set of one hop neighboring nodes with bidirectional links to act as multipoint relays (MPR). A MPR of node N is denoted MPR(N). MPRs are selected such that when N transmits a broadcast message, it is received by all nodes two hops away from N. Each node maintains information about the set of neighbors, termed MPR selector set, that have selected it as MPR. A node N informs another node that it has been selected as N's MPR using a periodic HELLO message. A node may modify its MPR over time and it indicates it in its HELLO messages.

MPRs are used in route calculation to establish a route from a given node to any destination. They are responsible for forwarding control traffic to disseminate information to the entire network and for announcing the link-state information periodically to determine the shortest path routes to their MPR selectors. A MPR announces that it is able to reach the nodes that selected it as their MPR. MPRs provide an efficient mechanism to control flooding by reducing redundant retransmissions in the same region. A neighbor of node N that is not its MPR receives and processes broadcast messages but does not retransmit the messages it receives from N.

6.3.1 Packet Format

A unified packet format is defined for all OLSR data. This approach facilitates extensibility without breaking backward compatibility. It also allows piggybacking of different types of information in a single packet. A packet encapsulates one or more messages (Figure 6.7).

The packet header contains the following fields:

- Packet length: The length of the packet in bytes.
- Packet SN: Incremented each time a new packet is transmitted.
- Message type: Indicates the type of message in the MESSAGE part. The valid value is between 0 and 127. The required message types for the core functions of OLSR are HELLO-messages (perform link sensing, neighbor detection, and MPR signaling), TC messages (topology declaration, i.e., advertisement

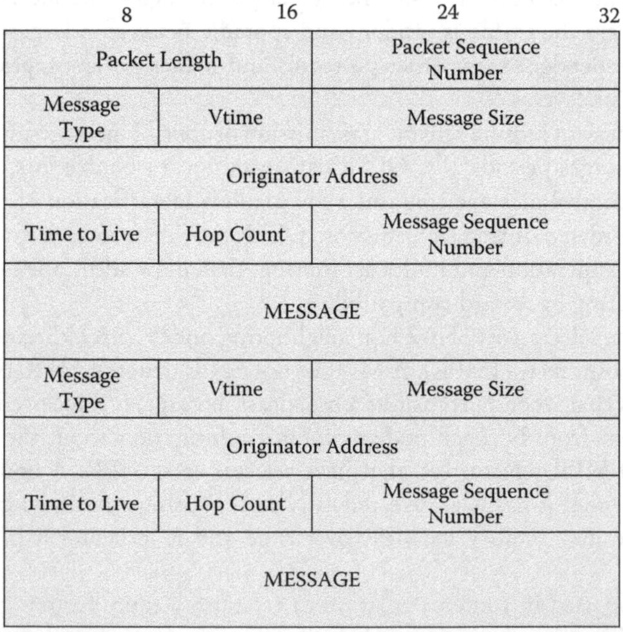

8	16	24	32
Packet Length		Packet Sequence Number	
Message Type	Vtime	Message Size	
Originator Address			
Time to Live	Hop Count	Message Sequence Number	
MESSAGE			
Message Type	Vtime	Message Size	
Originator Address			
Time to Live	Hop Count	Message Sequence Number	
MESSAGE			

Figure 6.7   OLSR packet format.

of link states), and multiple interface declaration (MID) messages (declare the presence of multiple interfaces on a node).

Vtime: Indicates for how long after reception a node must consider the information contained in the message as valid unless it receives a more recent update.

Message size: The size of this message in bytes, measured from the beginning of the message type field to the beginning of the next message type field or to the end of the packet (if there is no following message).

Originator address: The main address of the node that generates this message. It is unchanged in retransmission.

TTL: The maximum number of hops a message is transmitted. It is decremented by one before a message is retransmitted. If the value of this field is 0 or 1, the node that receives the message does not retransmit the message. This field allows an originator to limit the flooding radius.

Hop count: The number of hops a message has traveled. It is initially set to 0 and incremented by one before a message is retransmitted.

Message SN: The originator assigns a unique ID number to each message that is inserted into this field. The SN is increased by one for each message originating from the node.

OLSR packets are transmitted using UDP. A message can be flooded to the entire network or limited to nodes within a few hops from the originator.

6.3.2 Packet Processing and Message Flooding

Upon receiving a packet, the node examines the message type field to determine if the packet should be processed or discarded. Because a node may receive duplicates of the same packet, each node maintains a duplicate set that contains information about the most recently received packets to avoid reprocessing duplicate packets. If the duplicate set indicates the message has been processed, it is not processed again. Likewise, if the set shows that the message has been forwarded, it is not retransmitted and is discarded. A message is also discarded if:

The packet contains no messages: that is, the packet length is less than or equal to the size of the packet header.

The TTL is less than or equal to 0.

If the packet is not discarded, it is forwarded according to the default forwarding algorithm.

6.3.3 Information Repositories

Through the exchange of OLSR messages, each node accumulates information about the network. The information repository at each node contains:

■ The interface association tuples, (I_iface_addr, I_main_addr, I_time), for each destination in the network.
■ The logical link information base stores information about links to neighbors.
■ The neighborhood information base stores information about neighbors, two-hop neighbors, MPRs, and MPR selectors.
■ A node maintains a topology information base than contains topology information about the network. The information is acquired from TC messages and is used for routing table calculations.

6.3.4 HELLO Messages

The periodic exchange of HELLO messages is used to populate the local link information base and the neighborhood information base. The HELLO message is sent as the data portion of the packet format in Figure 6.7 with the message type field set to HELLO_MESSAGE. Figure 6.8 shows the format of a HELLO message. The fields of a HELLO message are:

■ Reserved: Set to 0.
■ Htime: Specifies the HELLO emission interval used by the node: the time before the transmission of the next HELLO.
■ Willingness: Specifies the willingness of a node to carry and forward traffic for other nodes. The value of this field is an integer between 0 and 7. A node that specifies a willingness of WILL_NEVER (0) must never be selected as MPR by any node. A node with willingness WILL_ALWAYS (7) must always be selected as MPR. The default value for this field is WILL_DEFAULT (3).
■ Link code: Specifies information about the link between the sender and the following list of neighbor interfaces. It also specifies information about the status of the neighbor.
■ Link message size: Specifies the size of the link message in bytes and is measured from the beginning of the link code field until the next link code field (or to the end of the message).
■ Neighbor Interface Address: Address of an interface of a neighbor node.

A HELLO message serves three tasks: link sensing, neighbor detection, and MPR selection signaling. The three tasks are based on periodic information exchange within a node's neighborhood for the purpose of local topology discovery.

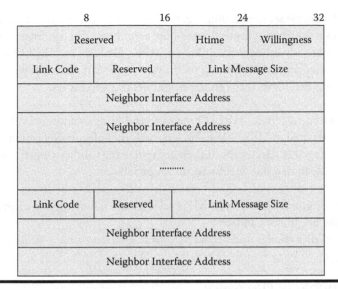

	8		16	24	32

Figure 6.8   Format of a HELLO message.

A HELLO message is generated based on the local link set, the neighbor set, and the MPR set in the local link information base.

A node performs link sensing on each interface to detect links between the interface and neighbor interfaces. A node advertises its entire symmetric one-hop neighborhood on each interface to perform neighbor detection. For a given interface, a HELLO message contains a list of links on that interface and a list of the entire neighborhood.

A HELLO message is broadcast by a node to its neighbors and is never forwarded by the recipients. The recipient of the HELLO message processes the message for the purposes of:

■ Link sensing: Concerned with populating the local link information base. For the purpose of link sensing, each neighbor node has a status of either symmetric (the link to the neighbor node is bidirectional) or asymmetric (communication from the neighbor node is possible but communication to the neighbor node is not confirmed). The information acquired by link sensing is accumulated in the link set.

■ Neighbor detection: Populates the neighborhood information base and is concerned with nodes and main addresses. A node maintains a set of neighbor tuples that is updated according to changes in the link set. There is a clear association between the link set that maintains information about the links and the neighbor set that maintains information about the neighbors. A node is a neighbor of another node if and only if there is at least one link between the two nodes.

■ MPR selector set population: Each node in the network selects its own set of MPRs among its symmetric one-hop neighbors. The symmetric links with MPRs are advertised with link type MPR_NEIGH in HELLO messages. A node calculates its MPR set in such a way that it can reach all symmetric strict two-hop neighbors through the neighbors in its MPR set.

When a node receives a HELLO message and finds its address in the list with a neighbor type equal to MPR_NEIGH, it records the information from the HELLO message in the MPR selector set. It then computes the validity time from the Vtime field. The node updates its MPR selector set as follows:

■ If the originator is not in its MPR selector set, a new tuple is created by adding the originator's address.
■ The validity time is updated as MS_time = current time + validity time.

6.3.5 Topology Discovery

The topology information is disseminated through the network via the mechanisms provided by link sensing, neighbor detection, and packet forwarding. These mechanisms provide an optimized flooding mechanism through MPRs. Routes are constructed through advertised links and links with neighbors. A node disseminates links between itself and the nodes in its MPR selector set to provide sufficient information to enable routing.

The format of a topology control (TC) message that is used for topology declaration is shown in Figure 6.9. This message is sent in the data portion of the packet shown in Figure 6.7 with the message type set to TC_MESSAGE. The TTL value is set to the maximum value (255) to disseminate the message to the entire network.

The fields in a TC message are:

■ Advertised neighbor SN (ANSN): A SN associated with the advertised neighbor set. Each time a node detects a change in its advertised neighbor set,

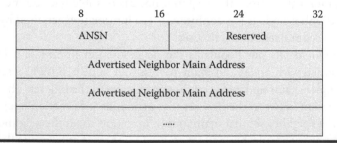

Figure 6.9　TC message format.

it increments the SN. It is used to keep track of the most recent information. When a node receives a TC message, it decides whether the information received is more recent than what it already has based on the ANSN.

■ Reserved: Set to 0.
■ Advertised neighbor main address: Contains the main address of a neighbor node. All main addresses of the advertised neighbors of the originator node are included in the TC message.

A node sends a TC message to declare a set of links (the advertised link set) that contains the links to all nodes in its MPR selector set. The ANSN number is incremented when links are removed from or added to the advertised neighbor set to build the topology information base, each node that has been selected as a MPR broadcasts TC messages that are flooded to all nodes in the network. The information disseminated by TC messages is used by other nodes to calculate its routing table.

6.3.6 Routing Table Calculation

Each node maintains a routing table that is used to route data destined for other nodes. Information in the routing table is based on the information in the local link information base and the topology set. Each entry in the routing table consists of R_dest_addr, R_next_addr, R_dist, and R_iface_addr. The entry means that the destination node, R_dest_addr, is R_dist hops away from the local node. The next hop node is R_next_addr and it is reachable through the local interface with the address R_iface_addr. The routing table is updated when:

■ A new neighbor appears or an existing neighbor is lost.
■ A two-hop tuple is created or removed.
■ A topology tuple is created or removed.
■ Multiple interface association information changes.

A routing table update does not generate any messages for transmission. A node runs a shortest path algorithm to construct its routing table.

6.4 TBRPF

TBRPF (Ogier et al. 2004) is a proactive, link-state routing protocol that provides hop-by-hop routing along the shortest paths. Each node computes a source tree using a modified Dijkstra's algorithm. It uses a combination of periodic and differential updates to keep neighbors informed of the reported part of its source tree. TBRPF consists of two modules: TBRPF neighbor discovery (TND) module and a routing module. TND is independent of the routing module. Because TBRPF does not require reliable nor sequenced message delivery, no acknowledgment is needed.

6.4.1 TBRPF Packets

A TBRPF packet consists of a header, optional header extensions, and a body that includes one or more messages and padding, if needed. A packet header (Figure 6.10) is of variable length with a minimum length of one octet. The fields in a header are:

- Version: Version number.
- Flag: Bits L and I specify which header extension follows. Bits R are reserved for future use and are set to 0.
- Reserved: For future use and is set to 0.

The TBPRF packet body (Figure 6.11) comprises a concatenation of one or more TBRPF messages and padding options. The fields are:

- Options: The value depends on type.
- Type: Identifies the message type or padding option.
- Value: Is of variable length.

There are two types of padding options (Figure 6.12) that may be inserted when necessary to align multioctet words within the TBRPF packet (modulo-8/4/2 addresses for 64/32/16-bit words, respectively). Pad1 inserts one octet of padding in the packet body and omits the value field. If more than one octet of padding is required, the PadN option is used. PadN inserts two or more octets of padding into the packet body. The first octet of the value field has a value between 0 and 253 that specifies the number of zero-value octets that immediately follow. The maximum length of padding is 255 octets.

The HELLO message (Figure 6.13) contains the following fields:

- HSEQ: A HELLO SN that is incremented by one each time a HELLO is sent.

Figure 6.10   TBRPF packet header.

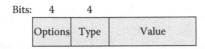

Figure 6.11   TBRPF packet body.

■ Pri: Takes a value between 0 and 15 that indicates the sending node's relay priority. A node with higher relay priority is more likely to be selected as the next hop on a route.

■ *n:* The number of neighbor interface addresses in the message.

A HELLO message has three subtypes: NEIGHBOR REQUEST (type = 2), NEIGHBOR REPLY (type = 3), and NEIGHBOR LOST (type = 4). HELLO is a concatenation of the three subtypes. The last two are omitted if the list of addresses is empty, but a HELLO message always consists a NEIGHBOR REQUEST, even if the list is empty.

A node sends a HELLO at least once per HELLO_INTERVAL (refer to Table 6.1). A node may send more than one HELLO during that interval, but the time between two HELLOs must be at least NBR_TIME/128 seconds. To avoid collisions of HELLO messages, the message is not transmitted at equal intervals. Instead, a node chooses an interval of (HELLO_INTERVAL—jitter) between consecutive HELLOs, where jitter is selected randomly between 0 and MAX_JITTER. The content of a HELLO from node I to node J is determined by the following three steps:

1. If nbr_status(I, J) is LOST and nbr_count(I, J) > 0, include J in NEIGHBOR LOST and decrement nbr_count(I, J).;
2. If nbr_status(I, J) is 1-WAY and nbr_count(I, J) > 0, include J in NEIGHBOR REQUEST and decrement nbr_count(I, J).
3. If nbr_status(I, J) is 2-WAY and nbr_count(I, J) > 0, include J in NEIGHBOR REPLY and decrement nbr_count(I, J).

When node I receives a HELLO on interface I from interface J of node J, it checks if an entry for J exists in its neighbor table. If an entry does not exist, one is created. Then the variables for the entry are updated accordingly.

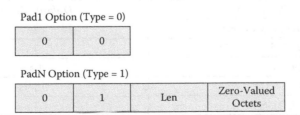

Figure 6.12   TBRPF padding options.

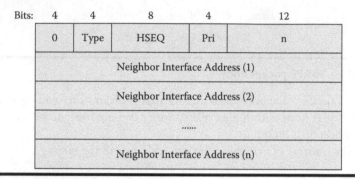

Figure 6.13 HELLO message format.

6.4.2 Neighbor Table

A node maintains a neighbor table for each interface. Table 6.2 lists variables maintained in a neighbor table. Assume that node i has an interface I, node j has an interface J, and (I, J) denotes a link between interface I and interface J.

6.4.3 TND

TND allows node i to detect a neighbor node j such that a bidirectional link (I, J) exists between interface I of node i and interface J of node j. It also detects when a bidirectional link breaks or becomes unidirectional. It uses differential HELLO messages that report only changes in the status of links, resulting in smaller sized HELLO messages compared to HELLO messages in OSPF. Consequently, HELLO messages can be sent more frequently to allow faster detection of topology change.

TND only senses neighbors that are one hop away. As with OSPF, nodes with multiple interfaces run TND on each interface, and a neighbor table is maintained for each interface. Because TND is designed to be modular and independent of the routing module, it can be used by other routing protocols. Likewise, TBRPF can use a different neighbor discovery protocol instead of TND.

Table 6.1 TBRPF parameters.	
Parameter	*Default Value*
HELLO_ACQUIRE_COUNT	2
HELLO_ACQUIRE_WINDOW	3
HELLO_INTERVAL	1 s
MAX_JITTER	0.1 s
NBR_HOLD_COUNT	3
NBR_HOLD_TIME	3 s

Table 6.2 Variables maintained in a neighbor table.	
Variable	*Description*
nbr_rid(I, J)	The router ID of the node associated with interface J.
nbr_status(I, J)	The current status of (I, J) which is either LOST, 1-WAY, or 2-WAY.
nbr_life(I, J)	The amount of time in seconds remaining before nbr_status(I, J) is updated to LOST if no further HELLO message is received from node j. The value is set to NBR_HOLD_TIME each time a HELLO is received from node j.
nbr_hseq(I, J)	The value of HSEQ in the last HELLO message received by node i from node j. It is used to determine the number of missed HELLO messages.
nbr_count(I, J)	The remaining number of times a NEIGHBOR REQUEST/REPLY/LOST message containing node j must be sent on node i.
hello_history(I, J)	A list of SNs of the last HELLO_ACQUIRE_WINDOW HELLO messages received on node i from node j.
nbr_metric(I, J)	An optional measure of the quality of the link (I, J). The valid value range is between 1 and 255, with smaller values indicating better quality. The default value of 1 means that this variable is not used.

Let us assume that node i has an interface I and node j has an interface J, and that (I, J) is a link between i and j. Node i maintains a neighbor table for each interface I to store information about each neighbor interface J heard on interface I. The status of a link is either 1-WAY, 2-WAY, or LOST.

A HELLO message is sent by a node to discover neighboring nodes. It contains a HSEQ that is incremented with each transmitted HELLO. A node sends at least one HELLO message per HELLO_INTERVAL. A HELLO message from node i contains three lists:

1. NEIGHBOR REQUEST contains a list of neighbor interface addresses that node i would like to establish a link with.
2. NEIGHBOR REPLY contains the interface addresses of neighbors that have listed an interface of node i in its NEIGHBOR REQUEST in their previous HELLO messages.
3. NEIGHBOR LOST contains a neighbor interface address J with whom link (I, J) is LOST.

It is possible that any of the lists may be empty. Whenever the status of a link (I, J) changes, node i includes node j's interface address in one of the above lists, as appropriate, in at most NBR_HOLD_COUNT consecutive HELLO messages sent on interface I. This ensures that one of the messages is either received on interface J or node j misses NBR_HOLD_COUNT HELLO messages. In the latter case, node j updates the status of link (J, I) to LOST.

To avoid establishing short-lived links, node i must receive at least HELLO_ACQUIRE_COUNT of the last HELLO_ACQUIRE_WINDOW HELLOs sent from interface J before declaring the link (I, J) to be 1-WAY. Then, node i includes J in the NEIGHBOR REQUEST list in the next NBR_COUNT HELLO messages sent on interface I or until a NEIGHBOR REPLY message containing node i is received on interface I from interface J. When node j receives the HELLO message on interface J, it declares link (J, I) to be 2-WAY. Then it includes node i in the NEIGHBOR REPLY list of the next NBR_HOLD_COUNT HELLO messages sent on interface J. Upon receiving the HELLO message on interface I, node i declares link (I, J) to be 2-WAY.

Node i updates the status of link (I, J) to be LOST if either:

- It receives a HELLO message on interface I from interface J with a HSEQ indicating that at least NBR_HOD_COUNT HELLOs were missed.
- It receives no HELLO message on interface I from interface J within NBR_HOLD_TIME seconds.

Then, it includes node j in the NEIGHBOR LOST list in the next NBR_HOLD_COUNT HELLO messages sent on interface I. Node J declares link (J, I) as LOST when it receives one of the messages or it misses NBR_HOLD_COUNT HELLO messages.

6.4.4 Routing Module

The routing module performs topology discovery and route computation. Each node maintains a topology table (TT) that stores information about each known node and link in a network. A node is identified by its router ID (RID), where node U is a node with RID U. A TT for node i stores information for each node U and link (U, V).

Each node maintains a source tree (T) that provides the shortest paths to all reachable nodes. A node computes T based on partial topology information stored in its TT using a modified Dijkstra's algorithm. A node reports only a part of its source tree, the reported tree (RT), to its neighbors to reduce overhead. The report is sent periodically. Periodic updates ensure that neighbors eventually learn of RT even if they do not receive all updates. Changes such as addition and deletion are reported in a more frequent differential update to ensure fast propagation of the update to all nodes affected by the change. A node does not forward a topology update message it receives.

There are three types of topology update messages: full, add, and delete. A full update reports that a set of links belongs to the sender's RT. An add update reports that a set of links has been added to the sender's RT. A delete update reports that a set of links has been deleted from the sender's RT. An update procedure, which is a variant of Dijkstra's algorithm, is executed periodically to update the source tree T and the topology graph (TG). The routing table is updated whenever there is a change to the source tree.

Each node generates and transmits a set of full topology update messages on all interfaces every PER_UPDATE_INTERVAL seconds. This is the periodic update. A differential update occurs every DIFF_UPDATE_INTERVAL seconds if it is not time to generate a periodic update but RT has changed since the last time a topology update was generated. In this case, a set of topology update messages that describes changes to the RT is generated and transmitted on all interfaces.

Each node checks for outdated topology information periodically based on the expiration timers. Expired entries are removed from the TG.

6.5   Summary

How well do the routing protocols work in a real-life implementation? Although there are several studies that have simulated the protocols to test their efficiency, currently there are very few real-life implementations of any of the protocols. Consequently, their efficiency and effectiveness is yet to be ascertained. Chin et al. (2002) attempt to address this concern by building a five-hop, four-node testbed deploying AODV. One of the problems they identified is the reliability of a route. Because AODV uses hop count to establish a route from a source to a destination, it does not take into consideration link stability. The transient radio links resulted in poor operation of AODV. To overcome this problem, they introduced a neighbor selection mechanism based on signal strength called powerwave. This mechanism is capable of distinguishing between stable and transient links, resulting in a route that is more stable.

There are a number of enhancements that have been proposed by other studies to enhance the operations of the routing protocols discussed above but are not discussed in this chapter. The studies are listed in the bibliography if you are interested to explore them further.

References

Chin, K.-W., J. Judge, A. Williams, and R. Kermode. 2002. Implementation experience with MANET routing protocols. *ACM SIGCOMM Computer Communications Review* 32(5):49.

Clausen, T., and P. Jacquet. 2003. Optimized link state routing protocol (OLSR). *IETF RFC3626.* Report. October.

Johnson, D. B., D. A. Maltz, and Y.-C. Hu. 2004. The Dynamic source routing protocol for mobile ad hoc networks (DSR). *IETF MANET Working Group Internet Draft.* Report. July.

Ogier, R., F. Templin, and M. Lewis. 2004. Topology dissemination based on reverse-path forwarding (TBRPF). *IETF RFC3654.* Report. February.

Perkins, C., and D. Das. 2003. Ad hoc on-demand distance vector routing (AODV). *IETF RFC3561,* Report. July.

Bibliography

Boukerche, A. 2004. Performance evaluation of routing protocols for ad hoc wireless networks. *Mobile Networks and Applications* 9(4):333.

DaSilva, L. A., S. F. Midkiff, J. S. Park, G. C. Hadjichristofi, J. D. Davis, K. S. Phanse, and T. Lin. 2004. Network mobility and protocol interoperability in ad hoc networks. *IEEE Communications Magazine* 42(11):88.

Kaba, J. T., and D. R. Raichle. 2001. Testbed on a desktop: Strategies and techniques to support multi-hop MANET routing protocol development. *Proceedings of the 2nd ACM International Symposium on Mobile Ad Hoc Networking and Computing (Mobi-HOC).* 164. New York: Association for Computing Machinery.

Lee, S. J., W. Su, and M. Gerla. 2002. On-demand multicast routing protocol in multihop wireless mobile networks. *Mobile Networks and Applications* 7(6):441.

Moghim, N., F. Hendessi. and N. Movehhedinia. 2002. An improvement on ad-hoc wireless network routing based on AODV. *Proceedings of the 8th International Conference on Communication Systems (ICCS 2002)* 2, 1068.

Perkins, C., and P. Bhagvat. 1998. Highly dynamic destination-sequenced distance-vector (DSDV) for mobile computers. ACM Computer Communications Review, *Proceedings of the Conference on Communications Architectures, Protocols and Applications (SIG-COMM '94)* **24**(4):234.

Zhong, X., S. Mei, Y. Wang, and J. Wang. 2003. Stable enhancement for AODV routing protocol. *Proceedings of the 14th IEEE Proceedings on Personal, Indoor and Mobile Radio Communications (PIMRC 2003)* 1, 201.

Online Resources

Internet Engineering Task Force (IETF). http://www.ietf.org/ (Accessed February 12, 2007).

Mobile Ad Hoc Network Research. http://cairo.cs.uiuc.edu/adhoc/ (Accessed February 12, 2007).

Publications on Mobile Ad Hoc Networks (MANETs). http://www.ee.surrey.ac.uk/Personal/G.Aggelou/ MANET_PUBLICATIONS.html (Accessed February 12, 2007).

Wireless Ad Hoc Networks Links. http://www.antd.nist.gov/wctg/manet/adhoclinks.html (Accessed February 12, 2007).

Chapter 7

Mobile IP

IP is the prevalent standard for communications on the Internet. An Internet address is used to route a packet from a source to a destination, possibly via intermediate routers. IP was designed at a time when computers remained at a fixed location. An Internet address implicitly specifies a computer's physical location. When delivering a packet to a particular destination, the same route is usually used unless an intermediate router falls out of favor due to congestion or other failures occur.

The proliferation of laptop and notebook computers coupled with mobility soon proved that IP is incapable of supporting seamless mobility. An IP address can no longer be used to determine the location of a destination node. Consequently, IP has to be extended to incorporate mobility support. The result is the Mobile IP specification (Perkins 1997, 1998), which was ratified by the IETF. This chapter discusses how Mobile IP supports mobility, how routing is handled, and how it addresses security concerns.

7.1 An Overview

A node may be a host or a router. A mobile node is a node that changes its point of attachment from one (sub)network to another without changing its IP address. A correspondent node is a mobile or stationary node that is communicating with a mobile node. A mobile node is associated to a home network, which has a network prefix matching that of the mobile node's home address. A home address is an IP address that is assigned to a mobile node for an extended period. The home address is unchanged regardless of where a mobile node is attached on the Internet. A foreign network is any network other than the home network. A home agent is a router on a mobile node's home network that delivers datagrams to mobile nodes in foreign networks. It maintains the current location information of each mobile node. A foreign agent is a router in a foreign network visited by the mobile node that cooperates with the home agent to complete datagram delivery to the mobile node. A mobility agent is either a home agent or a foreign agent. A visited network is network other than a mobile node's home network to which the mobile node is currently connected.

A mobility agent announces its existence by broadcasting agent advertisement messages. A mobile node that needs to find a mobility agent quickly and does not want to wait until the next advertisement message may multicast or broadcast a solicitation that is responded to by a mobility agent that receives it. A mobile agent determines whether it is on its home network upon receiving the advertisement message. If it is, it operates like any other node on its home network.

If the mobile node finds that it is on a foreign network, it obtains a care of address (COA) on the foreign network. A COA is the termination point of a tunnel toward a mobile node for datagrams forwarded to the mobile node while it is away from home. A tunnel is the path followed by an encapsulated datagram. While encapsulated, the datagram is routed to a knowledgeable agent that decapsulates it and forwards it to its ultimate destination. A mobile node obtains a COA by soliciting or listening for agent advertisement or contacting the DHCP or Point-to-Point Protocol. The mobile node registers its COA with its home agent via its foreign agent. Datagrams sent to the mobile node's home address are intercepted by its home agent, tunneled to the COA and received at the tunnel endpoint (either the foreign agent or the mobile node) and delivered to the mobile node. Datagrams sent by a mobile node from a foreign network are delivered to the destination using a standard IP routing mechanism, not necessarily via the home agent.

7.2   Agent Advertisement Message

A mobility agent broadcasts an advertisement message periodically. The mechanism used by a mobile node to detect a mobility agent is similar to that used to detect routers running the Internet Control Message Protocol (ICMP). ICMP was extended to support the needs of mobility agents. The extension format is illustrated in Figure 7.1. The type field is set to 3 to indicate that it is a mobility agent advertisement. The length field indicates the length of the packet and depends on how many COAs are advertised. The value of the SN field is incremented by one for each successive advertisement. The flags describe special features of the advertisement:

■ R: Registration with this foreign agent is required.
■ B: The foreign agent is busy.

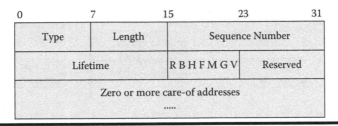

Figure 7.1 Mobility agent extension format.

H: This is a home agent.
F: This is a foreign agent.
M: Minimal encapsulation.
G: Generic router encapsulation (GRE)
V: Van Jacobson header compression.

Bits F and H are not mutually exclusive as a node may act as both a home and a foreign agent. Flag B can only be set if node F is also set. A busy foreign agent must continue broadcasting the advertisement message (with bit B set) because failure to do so would lead to mobile agents concluding that the foreign agent has crashed and is moving away unnecessarily. If a mobile node no longer detects advertisements from a foreign agent that previously offered a COA to the mobile node, the mobile node assumes that the foreign agent is no longer within range and begins searching for a new COA by waiting for an advertisement from another foreign agent or by sending a solicitation.

A home agent broadcasts an advertisement message without the COAs. The advertisement message broadcast by a home agent allows a mobile node to ascertain that it has returned to its home network.

In summary, an agent advertisement:

Allows for the detection of mobility agents
Lists one or more available COAs
Informs mobile nodes about special features that a foreign agent provides, such as alternative encapsulation techniques
Lets a mobile node determine the network number and the status of its link to the Internet
Lets a mobile node know whether the agent is a home agent, foreign agent, or both, and, in turn, determines whether it is in a home or a foreign network

7.3 Home Network Configurations

There are three possible configurations for home networks (Figure 7.2):

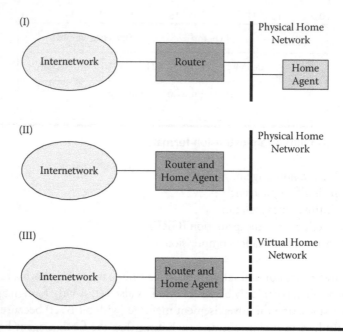

Figure 7.2 Home network configurations. (From Perkins, C. E. 1997. Mobile IP. *IEEE Communications Magazine* **25(5):84. Used with permission.)**

1. A standard physical network is connected by a router and another node on the network acts as a home agent. This is a common configuration for enterprises that are starting to deploy Mobile IP.
2. An alternative configuration is a physical network with the enterprise router doubling as a home agent.
3. A home network may be a virtual network. The home agent resides on the enterprise router and appears to the rest of the Internet as a router from the home network.

For the first two configurations, mobile devices can be configured with IP addresses of the existing physical network.

7.4 Registration Messages

There are two types of registration messages: registration requests and registration replies. Both are sent using UDP on port 434. The format of the registration messages are shown in Figure 7.3. A foreign agent is a passive agent in the registration procedure. It passes a registration request to the home agent and later passes the registration reply from the home agent to the mobile node.

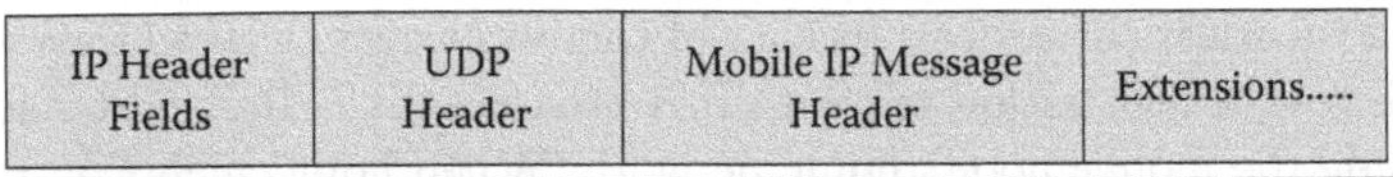

IP Header Fields	UDP Header	Mobile IP Message Header	Extensions.....

Figure 7.3 **Registration message format.**

7.4.1 *Registration Request*

A mobile node sends a registration request to inform its home agent of its current COA, how long it will be using the COA, and any special features that may be available from the foreign agent. The registration request is sent to the foreign agent that forwards it to the home agent. After the IP and UDP headers (Figure 7.3), the registration request has the format shown in Figure 7.4. The lifetime field specifies how long the mobile node intends to use the COA. A foreign agent may limit the registration lifetime to a configurable value that is indicated in its agent advertisements.

The flag bits M, G, and V are similar to the ones in the agent advertisement extension explained earlier. Bit B is used to inform the home agent to encapsulate broadcast datagrams from the home network for delivery to the mobile agent via its COA. Bit D indicates whether the mobile node is collocated with its COA and is used to determine how to deliver broadcast and multicast datagrams to the mobile node. The ID field is a security feature for replay protection.

When a home agent accepts the request, it associates the home address of the mobile node with its COA. This association is maintained until the registration lifetime expires. The triplet comprising the mobile node's home address, COA, and registration lifetime is termed as a binding for the mobile node. A registration request is considered a binding update.

If a mobile node is unable to contact its home agent, it may register with another unknown home agent using automatic home agent discovery. This is accomplished

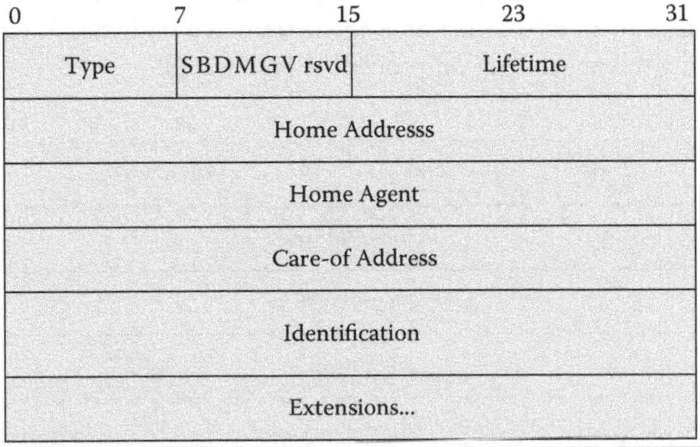

0	7	15	23	31

Type	S B D M G V rsvd	Lifetime
Home Addresss		
Home Agent		
Care-of Address		
Identification		
Extensions...		

Figure 7.4 **Registration request format.**

by using a broadcast IP address instead of the home agent's IP address as the target recipient of a registration request. This broadcast is a directed broadcast that reaches only the mobile node's home network. When home agents on the mobile node's home network receive the broadcast, they send a rejection containing their address to the mobile node. The mobile node may later use the address in the rejection notice to send a fresh registration request.

7.4.2 Registration Reply

The registration reply format is shown in Figure 7.5. The lifetime field indicates how long the registration will be honored by the home agent. It may be a shorter period than that specified in the lifetime field of the registration request, but never longer. If the registration fails, the code field is used to indicate the reason. If the value is 0, the registration is successful. A registration request may fail because it is denied by the foreign agent or the home agent. Table 7.1 shows possible values for the code field. The ID field is used by the foreign agent to match pending registration requests to registration replies and to relay the replies to the mobile node.

7.4.3 Secure Registration

An attacker may intercept datagrams meant for a mobile node by supplying a registration request with a bogus COA. Datagrams meant for the mobile node would then be sent to a different destination. To prevent such attacks, the registration process must be secure. There has to be a means to convince a home agent that a registration request really did come from the mobile node. The method used involves inserting a time stamp or a 32-bit random number, termed a nonce, in the ID field. This value changes with each new registration to avoid replay attacks. The home agent and mobile node have to be synchronized to use the time stamp or nonce. In case of an error, the home agent returns a registration reply with code value 133 and a resynchronization procedure is initiated.

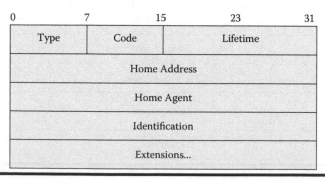

Figure 7.5   Registration reply format.

Table 7.1   Values of the code field.

Registration Denied by Foreign Agent		Registration Denied by Home Agent	
66	insufficient resources	130	insufficient resources
69	lifetime request > advertised limit	131	mobile node failed authentication
70	poorly formed request	133	registration ID mismatch
71	poorly formed reply	134	poorly formed request
88	home agent unreachable	136	unknown home agent address

The standard requires that a mobile node and home agent must share a security association and be able to use message digest 5 (MD5) with 128-bit keys to create unforgeable digital signatures for registration requests.

Mobile IP defines three authentication extensions: mobile-home authentication extension, mobile-foreign authentication extension, and foreign-home authentication extension. The mobile-home authentication is required in all registration requests and replies. The extensions have similar formats (Figure 7.6) and are distinguished by the value of the type field. The security parameter index (SPI) field defines the security context to compute and check the authenticator. The SPI selects the authentication algorithm and mode, as well as a secret to compute the authenticator. The secret may be a shared key or a public-private key pair. The default authentication algorithm is keyed MD5 in prefix and suffix mode to compute a 128-bit message digest of the registration message. The default authenticator is a 128-bit message digest computed by the default algorithm over the following three byte streams:

1. The shared secret defined by the mobility security association between the nodes and by SPI value specified in the authentication extension, followed by stream 2.
2. The protected fields from the registration message in the order specified above, followed by stream 3.
3. The shared secret again.

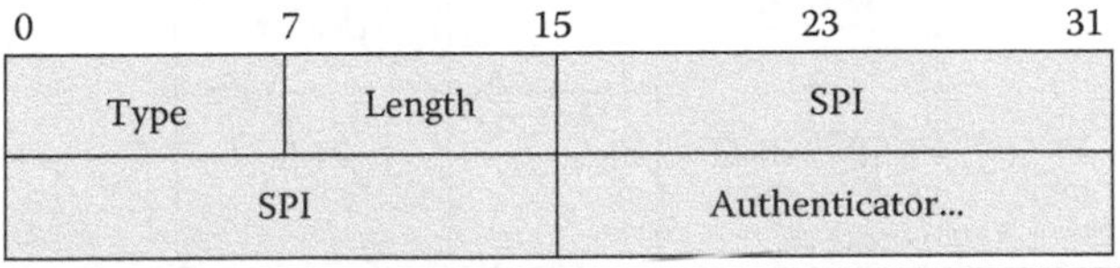

Figure 7.6   Authentication extensions.

The authenticator and the UDP header are not included in the computation of the default authenticator value. All implementations of Mobile IP are required to implement the default authentication algorithm.

7.5   Routing and Tunneling

After a successful registration, the home agent starts to intercept datagrams destined for the mobile node and tunnels them to the mobile node's COA. There are a number of encapsulation algorithms, but the default algorithm is the IP-within-IP encapsulation. Figure 7.7 illustrates how an IP-within-IP encapsulation is performed. An IP datagram is encapsulated by prefixing it with a new IP header (i.e., the tunnel header). The new IP header uses the mobile node's COA as the destination address. The new header indicates the presence of the encapsulated datagram by setting the value of the outer protocol field to 4. The old header is not modified except by incrementing the TTL value by 1.

Minimal encapsulation may be used if all parties involved—the mobile node, foreign agent, and home agent—agree to do so. Minimal encapsulation uses fewer bytes per datagram. The format of the encapsulated datagram is similar to IP-within-IP encapsulation. Figure 7.8 shows the minimal encapsulation format. Minimal encapsulation is indicated by setting the value of the protocol number in the encapsulating header to 55. The header length is either 8 or 12 bytes depending on whether the original source address is present. The S bit is set to 1 if the original source address field is present; otherwise, it is set to 0. The header checksum is a 16-bit 1's complement of the 1's complement sum of all 16-bit words in the minimal forwarding header. The original destination address is copied from the destination address field of the original IP header.

7.5.1   Soft Tunnel State

An ICMP error message returned to a home agent may not contain the IP address of the correspondent node that originates the tunneled datagram. Consequently, a home agent may not be able to notify the correspondent node that an error has

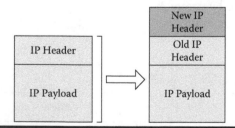

Figure 7.7   IP-within-IP encapsulation.

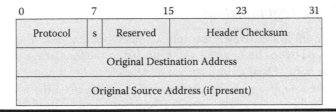

0	7		15	23	31
Protocol	s	Reserved		Header Checksum	
Original Destination Address					
Original Source Address (if present)					

Figure 7.8   Minimal encapsulation format.

occurred. To overcome this problem, a home agent keeps track of datagrams tunneled to a COA so that if it receives an ICMP error message, it is able to determine which datagram causes the error. In turn, this allows the home agent to relay the ICMP message to the correspondent node that sent the offending datagram.

For example, let us say a home agent is unable to forward a datagram to a COA due to a broken tunnel. Instead of returning a "network unreachable" ICMP error to the correspondent node, it returns a "host unreachable" ICMP error. This approach makes the tunnel transparent to the correspondent node, as it need not know that the mobile node is away from its home network.

In addition to keeping track of datagrams tunneled to a COA, a home agent also keeps track of other tunnel parameters (e.g., a path maximum transfer unit, a TTL for encapsulated datagrams using a tunnel). The collection of tunnel parameters is termed the soft state of the tunnel. Maintenance of the soft state helps in relaying ICMP messages.

7.5.2   *Proxy and Gratuitous Address Resolution Protocol*

If the home network is configured as a physical network (Figure 7.2), the home agent has to act as a proxy Address Resolution Protocol (ARP) for mobile nodes. Nodes on the home network that communicate with a mobile node are likely to have an ARP cache entry of the mobile node. When the mobile node leaves the home network, the cached entries become stale. For the nodes to be able to continue communicating with the mobile node, the home agent broadcasts gratuitous ARPs as soon as the mobile node registers a new COA. The gratuitous ARP causes the nodes attached to the home network to update their ARP caches so that they resolve the IP home address of the mobile node into the link layer address of the home agent. Likewise, when the mobile node returns to the home network, it broadcasts gratuitous ARPs so that its home address is once again associated to its own link layer address.

A mobile node is not allowed to broadcast an ARP request and reply packets to avoid creating irreparable stale ARP cache. For example, if a mobile node broadcasts an ARP request for its foreign agent's link layer address, stations on the foreign network may create an ARP cache entry for that mobile node. When the mobile node moves away, the cache entries become stale and there is no way to update it.

7.5.3 Route Optimization

When away from home, datagrams from a correspondent node to a mobile node are encapsulated by the home agent and tunneled to the foreign agent that relays it to the mobile node. Datagrams from a mobile node to other nodes can be routed directly via the foreign agent (there is no need to deliver it via the home network). What happens here is triangle routing (Figure 7.9) and is inefficient, especially when the correspondent node is closer to the visited network. This inefficiency is addressed using route optimization where a correspondent node has an up-to-date mobility binding. Mobility binding associates a home address with a COA for the remaining lifetime of that association. This binding information allows a correspondent node to encapsulate datagrams directly to the COA instead of relaying them via the home agent.

The home agent is responsible for providing binding updates to appropriate correspondent nodes in foreign networks. Mobility binding involves four steps:

1. A binding warning message (the format is shown in Figure 7.10) is sent to the home agent, indicating a correspondent node that is unaware of the mobile node's COA. It informs the recipient that the target node would benefit from obtaining a fresh binding for the mobile node.
2. If a correspondent node finds that its binding is stale, or going stale, it may send a binding request message (the format is shown in Figure 7.11) to the home agent. The 64-bit identification field is used to protect against replay attacks and to help match pending requests with subsequent requests.
3. The home agent sends an authenticated binding update message (the format is shown in Figure 7.12) containing the mobile node's current COA to correspondent nodes that need it. This event is triggered when a home agent receives a datagram destined for the mobile node from the correspondent node that has to be tunneled to the COA. A home agent need not wait for a binding warning or a binding request to send a binding update. The bind-

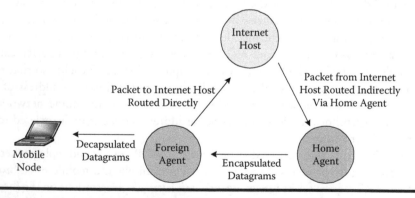

Figure 7.9 Triangle routing. (From Perkins, C. E. 1997. Mobile IP. *IEEE Communications Magazine* 25(5):84. Used with permission.)

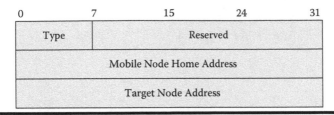

Figure 7.10 Binding warning message format.

ing update includes a lifetime for the COA association. The correspondent node purges the binding after the lifetime expires. The home agent informs the correspondent node that it may use minimal encapsulation or GRE to tunnel datagrams to the mobile node by setting the M or G flag bit, respectively. The A bit is set if an acknowledgment is required while the I bit is set if the ID field is present. A binding update must be accompanied by the route optimization authentication optimization that is similar to the mobile-home authentication extension.

4. The binding acknowledgment message (the format is shown in Figure 7.13) is sent to acknowledge a binding update message. An acknowledgment is required for a smooth handoff. Once again, the ID field protects against replay attacks and allows an acknowledgment to be matched with a pending binding update. If the N bit is set, it serves as a negative acknowledgment to inform the home agent that the update is not acceptable.

The messages are sent using UDP. All reserved fields are set to 0 and ignored upon reception.

7.5.4 Handoff Procedure

A handoff occurs when a mobile node moves from one point of attachment to another. The transition should be as smooth as possible. When a mobile node leaves a point of attachment, it is possible that datagrams tunneled to the previous point of attachment would be dropped because it is impossible to instantaneously inform all correspondent nodes of the change in the point of attachment. Route optimization addresses this problem by allowing a foreign agent to maintain a binding for their former mobile

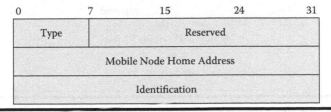

Figure 7.11 Binding request message format.

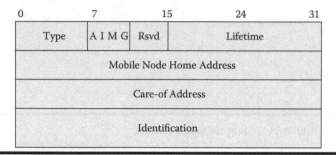

Figure 7.12 Binding update message format.

visitors by maintaining the current COA. Any datagrams destined for the mobile node are reencapsulated by the former foreign agent and relayed to the new foreign agent.

Because a smooth handoff aims to handle datagrams without dropping them while the mobile node is in transition between points of attachment, it is best not to involve the home agent as it is often too far away for a timely response. Figure 7.14 illustrates the process involved to facilitate a smooth handoff. When a mobile node moves to a new point of attachment, it instructs the new foreign agent to send a binding update to its previous foreign agent.

If the previous foreign agent does not have a fresh binding for the mobile node, it returns the datagrams to the home agent by decapsulating the datagrams and sending them using normal IP routing. The problem with this approach is it may give rise to a routing loop if the foreign agent loses track of the visiting mobile node. Route optimization overcomes this problem by defining special tunnels that indicate to the home agent that special handling is required. Instead of returning the decapsulated datagram using normal IP routing, the foreign agent encapsulates the datagrams using the foreign agent's COA as the source IP address. When the home agent receives the datagram, it compares the source IP address with the COA in the binding of the last registration. If the addresses match, the home agent understands that it must not tunnel the datagram to the COA. Otherwise, the home agent is allowed to retunnel the decapsulated datagrams to the current COA from the registration.

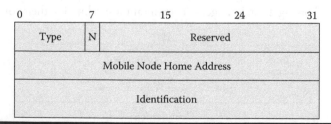

Figure 7.13 Binding acknowledgment message format.

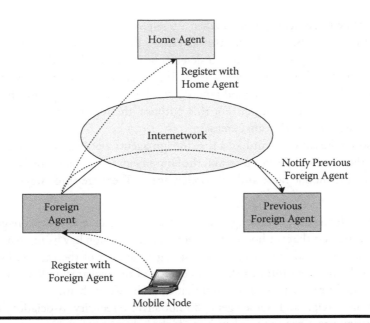

Figure 7.14 Smooth handoff during registration. (From Perkins, C. E. 1997. Mobile IP. *IEEE Communications Magazine* 25(5):84. Used with permission.)

7.6 Security Issues in Mobile IP

It is imperative that a malicious entity is not allowed to disguise itself as a valid user to gain access to network resources. The registration procedure addresses this concern with the use of a registration key. Additionally, when a mobile user roams in a foreign network, authentication is required not only to verify the identity of the user but may also be required for billing purposes. This concern is addressed by introducing an accounting, authentication, and authorization services in Mobile IP.

7.6.1 Security of the Registration Procedure

During a binding update, a mobile node has to convince the previous foreign agent that the binding update is not forged by showing the foreign agent that it possesses a registration key. The procedure to obtain a registration key is as follows:

■ The foreign agent uses agent advertisement flags and extensions to indicate what security features it offers.
■ The mobile node selects one of the available features.
■ The foreign agent responds to the mobile node's request and, if necessary, cooperates with the mobile node to provide smooth handoff and to obtain a registration key from the home agent.

A mobile node determines whether a foreign agent is willing to take part in a smooth handoff by inspecting the advertised flags. When a mobile node first detects a foreign agent, it is able to determine immediately if the foreign agent supports the mobility security association. If it does, the mobile node establishes a registration key by picking a good random number and encoding it for the foreign agent using their shared secret. For this purpose, the registration must include a mobile-foreign authentication extension.

However, because it is unlikely that foreign agents would support this security feature, the mobile node has to rely on the home agent to generate a registration key for use by the mobile node and the foreign agent. There are three ways to accomplish this:

1. If the foreign and home agents share a security association, the foreign agent may request that the home agent encrypt a registration key using that security association and send it to the foreign agent as part of the registration reply. The home agent informs the mobile node of the registration key value using the mobility security association that exists between them.
2. If the foreign and home agents do not share a security association, but the foreign agent has a public key, it sends the public key to the home agent along with the registration and requests that the registration key be encrypted using the public key. The encrypted value is returned as part of the registration reply.
3. If the foreign agent does not have a security association with the mobile node or the home agent, and it does not have a public key, a Diffie–Hellman key exchange may be executed.

7.6.2 Mobile IP and Accounting, Authentication, and Authorization

Mobile IP was initially designed to work in WLANs. To support roaming users, Mobile IP has been extended to perform accounting, authentication and authorization (AAA) services (Perkins 1999, 2000). The combination of Mobile IP and AAA provides the mobility infrastructure that meets the needs of cellular telephone for mobile data Voice-over IP and a large population of mobile telephone users. For a user to be billed for accessing and utilizing resources in a visited network, the identity of the user must be verified. This is where AAA plays an important role. Before we explain the implementation of AAA in Mobile IP, we will give an overview of the basic model of implementing AAA on the Internet.

An attendant is an agent in a foreign domain that attends to a client's request. When a client requests access to resources, the attendant requires that the client present credentials to authenticate itself before access is granted. To do so, the following have to be in place:

■ An attendant has to consult a local authority in the same foreign domain to obtain proof that the client's credentials are acceptable. Because the attendant and the local authority are in the same domain, they should have security associations that allow them to transact information securely.

■ The local authority may not have all the information required to verify the client's credentials, but it is expected that it is configured to negotiate the verification of the credentials with an external authority.

■ The local and external authorities have sufficient security associations and access control to negotiate the authorization to allow the client access the requested resources. This authorization depends on secure authentication of the client's credentials.

Figure 7.15 depicts the security relationships between the involved parties. Once the local authority obtains the authorization and the attendant is notified, the attendant may provide the user with access to the requested resources. Additional requirements for an AAA server are that it has to be able to obtain an IP address for the client, and it has to be able to identify the client by means other than its IP address.

Because there may be many attendants in a domain to handle many clients from different home domains, an attendant should run on inexpensive equipment. Because an attendant may serve a number of clients at a time, it has to keep the state of pending clients' requests while the local authority contacts the external authority. Attendants should be able to handle any QoS requests that accompany a client's request. Credentials that grant a client access to the requested resources should not be usable in future requests with the same or other attendants. To protect attendants from replay attacks, intermediate nodes must not be able to learn any information that may enable them to reconstruct the credentials to avoid forgery.

The basic model above is modified slightly (Figure 7.16) to incorporate AAA in Mobile IP. The attendant is the foreign agent. The home agent plays a role that is

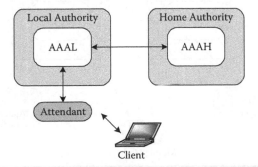

Figure 7.15   Basic model: AAA servers in home and local domains. (From Perkins, C. E. 2000. Mobile IP joins forces with AAA. *IEEE Personal Communications* **7(4):59. Used with permission.)**

subordinate to the role played by AAAH (accounting, authentication, and authorization home) during the initial registration. The initial registration with AAA takes slightly longer than subsequent registrations. After the initial registration, the mobile node is authorized to continue using Mobile IP at the foreign domain without further involvement of the AAA servers. The AAA servers perform the following additional tasks:

■ Initiate and enable the authentication for Mobile IP registration.
■ Once the identity of the mobile node has been established, they grant authorization for the mobile node to use Mobile IP and specific services.
■ Initiate the accounting for service utilization.

Referring to Figure 7.16, the security associations involved are:

■ SA_1: Between the user and his home domain.
■ SA_2: Between the home agent and AAAH. It serves to minimize the configuration burden of the mobile node because it allows the mobile node to maintain only one security association with AAAH and with no other.
■ SA_3: Between AAAL and AAAH and is needed to guarantee the reliability of the AAA processes. Nodes in different administrative domains, such as AAAH and AAAL, must be able to verify the identity of the party they are communicating with, and if required, guarantee the privacy of data.
■ SA_4: Between the foreign agent (attendant) and AAAL so that the foreign agent knows that the mobile node has been granted access to the local resources.

Mobile IP data sent by the foreign agent through AAAL and AAAH is opaque to the AAA servers. Authorization data needed by the servers are supplied by the for-

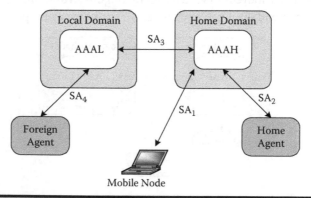

Figure 7.16   AAA servers with Mobile IP agents. (From Perkins, C. E. 2000. Mobile IP joins forces with AAA. *IEEE Personal Communications* **7(4):59. Used with permission.)**

eign agent from the data supplied by the mobile node. Hence, the foreign agent acts as a translation agent between the registration protocol and AAA. The AAA servers have to be able to perform key distribution during the initial registration from any administrative domain. The key distribution offers the following functions:

- Identifies or establishes a security association between a mobile node and a home agent that is needed for Mobile IP to work.
- Identifies or establishes a security association between the mobile node and foreign agent. The association is used by the same foreign agent as an assurance that the same mobile node is requesting continued authorization for Mobile IP services.
- Identifies or establishes a security association between the home and foreign agents. The association is used with subsequent registrations at the same foreign agent so that the foreign agent can be assured that the same home agent has continued the authorization of Mobile IP services for the mobile node.
- Participates in the distribution of the security association to the Mobile IP entities.
- The AAA server has to validate certificates provided by the mobile node and provides reliable indication to the foreign agent.

The lifetime of associations distributed by the AAA server should be long enough to avoid frequent initiation of AAA key distribution to avoid long delays between registrations. Long delays may result in dropped packets and noticeable service disruption.

7.7   Mobile IP and Ad Hoc Networks

The importance of ad hoc networks was discussed in Chapter 6. The characteristics of an ad hoc network were not considered during the initial design stage of Mobile IP. Therefore, a few modifications were made for Mobile IP to support ad hoc networking (Perkins 1996). The initial Mobile IP specification assumed that a mobile node has a direct link to a foreign agent. For a mobile node to use Mobile IP, it has to register with its home agent and obtain a COA from a foreign agent. In an ad hoc network, it is possible for a mobile node to stray away from the range of a foreign agent. Mobile IP has been extended to allow a mobile node to obtain a COA even if it is more than one hop away from a foreign agent. Conversely, foreign agents are modified to use an ad hoc route to a mobile node instead of forwarding packets on a directly connected link.

Another modification is the addition of a callback feature in the Mobile IP daemon. The callback feature allows applications running on the mobile node to request that they be informed when a mobile node's point of attachment changes. This is an important feature because when the point of attachment changes, environmental factors that affect the applications may also change: for example, available bandwidth and the cost of transmitting packets. Each time the point of attachment changes, the daemon sends a callback notification, which includes the new COA, to the applications.

When a mobile node roams in a foreign network, it is unaware of what resources are available. For example, how does a mobile node know if a printer is available to send a file for printing? To determine if a printer (or any other resource) is within the vicinity, the mobile node needs the assistance of a Resource Discovery Protocol (RDP). RDP allows a service (e.g., a printer, fax, file server) to register with a local directory agent. User agents that reside on a mobile node may contact the directory agent to obtain the URL that points to the required service. In addition to a URL, the directory agent also stores relevant information about a resource such as a color laser jet printer. Clients of RDP can name the service and specify the details about the service they require to locate a printer.

A nomadic-aware application that performs resource discovery posts a callback request to the Mobile IP daemon. When a mobile node moves, the application is notified of the move and is informed of the new COA. This ensures that the application is aware of the fact that the printer it used previously is no longer available and that if it needs a printer, it has to contact a directory agent to obtain information about a printer in its new domain.

7.8 Summary

Mobile IP was designed with the objective of supporting mobility. Much work has been done to provide efficient communication and routing to support mobile users. Security is one of the critical issues in supporting mobility. Not only do the users and the networks have to be protected from malicious entities that may attempt to intercept communications, it is also imperative that a user's privacy is protected. A user's location and movement information should not be made available to any third party as this information may be manipulated for illegal purposes, such as stalking.

Mobile IP has also been enhanced to support ad hoc networks. This enhancement provides more flexibility for Mobile IP to support various forms of mobile activities. Not only can Mobile IP be used in an infrastructured wireless network, but also in an infrastructureless environment. With this feature, Mobile IP may serve as another means to support seamless mobility.

References

Perkins, C. E. 1996. Mobile IP, ad-hoc networking, and nomadicity. Proceedings of the 20th International Computer Software and Applications Conference (COMPSAC '96):472–476.

Perkins, C. E. 1997. Mobile IP. *IEEE Communications Magazine* 25(5):84.
Perkins, C. E. 1998. Mobile networking through Mobile IP. *IEEE Internet Computing* 2(1):58.
Perkins, C. E. 1999. Mobile IP and security issues: An overview. *Proceedings of the 1st Internet Technologies and Services*:131–148.
Perkins, C. E. 2000. Mobile IP joins forces with AAA. *IEEE Personal Communications* 7(4):59.

Bibliography

Perkins, C. E. 1996. IP mobility support. *IETF RFC2002*. Report. October.
Perkins, C. E. 1998. Application of Mobile IP to tactical mobile internetworking. *Proceedings of the IEEE Military Communications Conference (MILCOM 1998)* 2:408–414.

Online Resources

Dynamics Mobile IP. http://dynamics.sourceforge.net/ (Accessed February 14, 2007).
Mobile IP: Connecting the World. http://www.acm.org/crossroads/xrds7–2/mobileip.html (Accessed February 14, 2007).
MobileIP.org. http://www.mobileip.org/ (Accessed February 14, 2007).
Wireless networking and Mobile IP references. http://www.cse.wustl.edu/~jain/refs/wir_refs.htm (Accessed February 14, 2007).

Chapter 8

Issues in Mobile Computing

There are several characteristics of wireless networks and mobile devices that operate in such environments that give rise to several issues that do not pose a problem in a fixed, wired infrastructure. Wireless networks are less reliable, have limited bandwidth, and transmissions are more susceptible to errors and easier to intercept. Mobile devices are often small and have limited processing capabilities. Mobile applications have to take into consideration the constraint of small displays, and heavy-duty processing should not be done on mobile devices. Because mobile devices operate on batteries, managing energy consumption economically becomes imperative to prolong operating time and reduces the frequency of recharge between use. Among the constraints that have to be considered are:

 Limitations on local resources, such as bandwidth and disk space.
 Adaptive behavior; that is, the ability to adapt to its environment and user demand.
 Battery power consumption.
 Heterogeneous environment (e.g., platforms, devices, underlying networks).
 Small display size gives rise to the need for new interface design techniques.
 Disparity in the availability of remote services.
 Unpredictable variation in QoS.

This chapter discusses various constraints of mobile computing and communications, how these constraints are addressed, and issues that are still outstanding.

8.1   Bandwidth

Bandwidth is one of the most critical resources in wireless networks. Even though offered data rates have increased significantly with the introduction of 3G, the data rates on wireless networks will always be a few magnitudes lower than what is available on wired networks. Design of mobile applications must take this discrepancy into account.

An issue that must be kept in mind when designing a mobile application is intermittent connectivity. Consider this scenario: a user in a train is using a mobile banking application to transfer funds. While this transaction is executing, the train enters a tunnel and the connection is lost. How does the application handle this lost of connectivity? It is unacceptable if the lost of connectivity results in a situation where money is deducted from one account, but not deposited in another. Ideally, when the application detects the loss of connectivity, it should be able to suspend the transaction and resume it when the train leaves the tunnel and it is able to reconnect to the server.

Mobile applications must be designed to adapt to fluctuations in available bandwidth, as it is an important resource that may have a huge impact on user experience and perceived usability of an application. If the connection is frequently lost and the user has to repeatedly restart a transaction, this can result in frustration and the user may regard the application as useless and cease to use it. Adapting to available bandwidth is an important aspect of adaptive behavior that must be considered when designing mobile applications.

8.2   Adaptive Behavior

Adaptive behavior arises due to a significant mismatch between supply and demand for resources. The mismatch often occurs in low-level system resources such as bandwidth, battery power, or memory. It may also relate to interactions such as display size or input modality. As a mobile user moves, services and resources may become available or disappear. As this happens, the mobile system needs to adapt to what is available in its environment.

The mismatch may occur even if the user is static. Available bandwidth varies not only because coverage is nonuniform over an area, but also due to the actions of neighboring users. For example, as more users move into a cell and compete for resources, the perceived QoS may decrease. As this happens, the system has to handle the variations in ways that minimize inconvenience to the users and do not abruptly interfere with the task the users are trying to accomplish. As an example, a few options may be offered to a wireless video conferencing application to adapt itself: The system may reduce the fidelity of the video, try to find another higher bandwidth connection, or inform the user of the bandwidth constraint and drop the video. Which action should be taken depends on what task the user is trying to accomplish.

In addition to user mobility, there are a few other reasons for a mismatch. The availability of computing servers or data-staging servers is location-dependent and affects techniques such as remote execution and cyber foraging. In the case of mobile code, low-level resources and interactive resources may vary widely between the source and destination systems. As time elapses, the resource level may change; for example, residual power on a laptop is depleted as the battery drains.

The mismatch cannot be ignored as it will result in an unsatisfactory user experience. Severe performance degradation may frustrate the user and, in turn, may result in increased human error. Adaptation is important not only as users move from a resource-rich to a resource-poor environment, but also in the other direction. For example, a user should not be deprived of speedier file transfer if more bandwidth becomes available.

Adaptation decisions require timely knowledge of current resource levels. The change in available resources and its detection should occur promptly. If the change justifies a modification in the amount of resources consumed by an application, it should be executed without delay. For this task to be accomplished efficiently, the operating system has to play a role in managing resources, such as bandwidth and battery power, in addition to its traditional role in managing resources such as CPU cycles and memory. Adaptation is necessary when there is a significant mismatch between supply and demand of resources. There are three adaptation strategies:

1. A client guides the application to use less of a scarce resource. This usually changes the user-perceived quality (also termed fidelity); for example, reducing the quality of video.
2. A client requests a guaranteed amount of resources. This involves a reservation-based mechanism. To be viable, this approach must provide incentives for resource providers to offer differential treatment of users.
3. A client suggests that the user take a corrective action. If the user agrees, resource supply may be adequate to meet the demand, but there is no guarantee.

All three strategies are important and which strategy is chosen depends on the smart space environment. Depending on the circumstances, the system may need to switch between the different strategies to guarantee a smooth and seamless transition. The first strategy is desirable because it hides the complexity of monitoring the available services and resources, but it is not always achievable. Although the second strategy seems preferable because it shields the user from reduced fidelity and taking corrective actions, it might not be suitable under all circumstances and in all environments. Reserving resources may deprive other users from obtaining the QoS that they require. Some form of priority mechanism may be required to determine who can reserve resources and which resources can be reserved, resulting in increased system complexity. The third strategy needs to be designed in such a way that it does not cause a distraction or becomes an annoyance to the user. For this purpose, a nonintrusive API is required.

There has to be a balance between providing a proactive system and transparency. A proactive system has to be carefully designed to avoid annoying the users, thus, defeating the goal of invisibility. As a system adapts to a changing environment, it has to take into consideration user preferences.

Odyssey (Noble and Satyanarayanan 1999) is a platform for mobile data access that supports application-aware adaptation that is targeted at applications that access rich, resource-demanding data. It defines two important attributes to support the adaptability feature in mobile systems, namely agility and fidelity. Agility is the speed and accuracy with which an application detects and responds to changes in resource availability. Data accessed by an application may be stored at a few locations, for example, file servers, SQL servers, or video libraries. Ideally, data presented to a mobile client should be identical to the original copy stored at the server, but this may be difficult to achieve due to resource constraints. Agility is a complex attribute because different applications may have different sensitivity to different resources. For example, a mobile conferencing system is sensitive to bandwidth fluctuation. It is more concerned with providing an acceptable QoS as bandwidth fluctuates. On the other hand, an application that involves intensive computation is more concerned with power drain as the CPU consumes a significant amount of power. In this case, the concern is to offload computation to a fixed network. These two applications require different mechanisms to detect different types of changes; therefore, different factors need to be considered in providing an agile application.

Fidelity is the degree to which data presented to a mobile client matches the reference copy at the server. A reference copy is the most complete, current, and detailed version of a data item. There are a number of factors that determine fidelity, depending on the type of data and the applications using it. For example, file systems such as Coda may deliver stale data to an application when network connectivity is poor or unavailable. On the other hand, a topography map has a dimension of minimum feature size and resolution. When bandwidth is scarce, an application may choose to download a map with a lower resolution to reduce the demand on limited bandwidth.

These examples illustrate the dissimilarity between different data types. To support adaptability for efficient resource usage, some form of awareness has to be incorporated into the system. For instance, a video data packet received in error can be dropped instead of retransmitted, but a database update must be received accurately and retransmitted if an error is detected. Odyssey's approach to support adaptability trades off data quality for resource consumption. Hence, when bandwidth drops, a client playing a full-color video may switch to a black-and-white video. When implementing the strategies to support adaptability, we have to keep in mind how fidelity for a broad range of applications can be lowered without causing too much inconvenience to the users.

The responsibility of adaptation can be made to be the sole responsibility of an application without the involvement of the operating system. This approach is termed laissez-faire adaptation, which is where each application adapts according to its own goals, disregarding the needs of other applications. Consequently, each application adapts to the same set of environmental changes and competes for the same scarce resources. As applications are more likely to interfere with one another, laissez-faire adaptation cannot support concurrency. Due to this drawback, Odyssey takes an approach where the operating system is responsible for monitoring resource availability, notifying applications of any changes, and enforcing resource allocation decisions. In addition to providing overall efficiency, this approach also supports concurrency. An approach that makes the system responsible for adaptation is termed application-transparent adaptation. Among resources monitored are bandwidth, latency, disk space, computing power, remaining battery power, and cost (in cases where the service used charges a fee).

The Odyssey architecture is comprised of:

- A warden: A set of type-aware code components. There is one warden for each type. It takes advantage of type knowledge to optimize resource usage and consistency. It provides fidelity options for applications to choose. The figure shows wardens for video and Web data types. Each warden may perform different operations to change data fidelity as appropriate.
- An interceptor: Provides a client that forwards file system requests on Odyssey objects to the viceroy.
- A viceroy: Responsible for all type-independent functionality on the client. It stores information regarding resource usage of all applications and monitors the availability of resources and manages their usage. It acts as a filter by notifying an application of a change in resource availability only if the change affects the application.

There are two system calls, a resource request and a type-specific operation, that perform adaptation functions. A resource request informs Odyssey of changes in resource availability; for example, the amount bandwidth has dropped and that the video player application needs to adapt to the change accordingly. Only relevant changes in available resources are relayed to an application, and only if the change has a significant impact on QoS. If the amount of change in the level of resources is small and does not have an adverse impact on the application, the information is not relayed. An application's tolerance toward changes in available resources is defined by a range termed window of tolerance. Once an application has chosen a fidelity level, it informs the system of its window of tolerance by issuing a resource request. The resource request is forwarded to the viceroy, which records it. If an estimate of resource availability is outside the range defined by the window, an upcall (type-specific operation) is issued to notify applications affected by the change; for example, the video player reduces its frame rate when bandwidth drops below a certain value. The application responds to the notification by changing the data fidelity. The task of changing the fidelity is carried out by the warden.

Adaptability is especially important to support a ubiquitous computing environment. Weiser's (1991) original vision for ubiquitous computing, now often referred to as pervasive computing, is that computers should disappear into the background. One day, computers will "weave themselves into the fabric of everyday life" until they are indistinguishable from it. The computing and communication capabilities are gracefully integrated with users' environments so that the technology "disappears." It is a technology that supports mobility so that users would not detect its absence as they move. For computers to disappear and be unobtrusive, computers need to have the ability to sense a user's context and environment. Thus, adaptability is critical to support pervasive computing. In addition to the issues that have been discussed, pervasive computing extends the problem domain as follows:

- Effective use of smart space: A space may be an indoor space (e.g., a meeting room) or a well-defined open area (e.g., a tennis court). Embedding computing infrastructure in a building infrastructure enables the bringing together of two worlds that were previously disjointed and allows one world to control another. For example, the air-conditioning and lighting in your office is turned on according to your preference when the sensor detects that you are heading toward your office from the parking lot.
- Invisibility: Weiser's idealistic vision is to have the technology disappear completely from a user's consciousness. A more reasonable approximation is to minimize user distraction. The technology should continuously meet the user's expectations without surprises. For example, as available bandwidth drops, the video conferencing application offers black-and-white video instead of shutting down.
- Localized scalability: As the interaction between users and their surroundings increases, it gives rise to bandwidth, energy, and distraction implications. This is compounded as the number of users sharing the space increases. Local scalability is critical to make efficient use of local resources, such as to reduce communication traffic and, hence, better utilize available bandwidth. The interaction between you and your office space, for example, has to decrease as you move away from it. A good system design should significantly reduce interactions between distant entities.
- Masking uneven conditioning: The level of smartness of available technology will likely vary in different environments and this will be noticeable to users. To reduce the amount of variation to the user, the computing space has to compensate for "dumb" environments. For example, a system that is incapable of a disconnected operation tries to mask the absence of wireless coverage in its environment by displaying a friendly error message.

To provide effective proactivity, a pervasive computing system should track a user's intent. Consider a video conferencing application. A conventional application for a fixed network assumes that a user has a fixed bandwidth allocated to it and makes no attempt to adapt itself. As available bandwidth in a wireless network fluctuates, a wireless video conferencing application has to adapt itself. There are a number of options: the system may reduce the fidelity of the video, it may try to find another higher bandwidth connection, or it may inform the user that the task cannot be accommodated and drop the video. Which action should be taken would depend on what task the user is trying to accomplish. For the system to be able to take the correct action without distracting the user, it has to be able to determine user intent, which is a nontrivial task. The following questions must be answered to determine user intent:

■ Can user intent be inferred, or does the user have to explicitly state intent? In the latter case, can it be statically specified (e.g., manual configuration in the system setting) or is it obtained through a series of interactions with the user?

■ How is user intent represented internally, and how rich must it be for it to be useful? As user intent may vary depending on the task the user tries to accomplish, how is the information updated? How is this information made available to different layers of the system?

■ How do we determine the accuracy of knowledge of user intent? Is incomplete or imprecise knowledge still useful? At what level of uncertainty is it better to ignore it when making decisions?

■ Will obtaining the intent cause inconvenience to the user or affect usability and performance? How does one trade off between the benefit and the overhead cost of obtaining and maintaining the information?

Addressing these concerns is nontrivial and is a focus of ongoing research.

As technology advances and more services are available, users may require computing power that poses a challenge to providing other mobility requirements, such as small, light devices with long battery lifetimes. An approach to address this problem is cyber foraging, where the computing power of mobile devices is augmented by using machines on the wired infrastructure (Satyanarayanan 2001). As the price of computers continues to drop, it will be possible to provide computing servers or data staging servers that are connected to the Internet at public spaces such as train stations and libraries. These servers are surrogates, and act as gateways to the Internet. When a mobile device enters a new neighborhood, it tries to detect the presence of surrogates and negotiates the terms for the use of their services. If intensive computation is required, the computation is migrated to the surrogate. The surrogate may retrieve data from the Internet and cache it locally while performing the computation. The system may also be configured in such a way that the arrival of the user in a neighborhood is anticipated and data required by the user is retrieved in advance. Caching data in advance reduces the delay perceived by the user. When the user leaves the neighborhood, any data staged or cached on his behalf is discarded.

A number of issues need to be addressed to support cyber foraging:

- How does one detect the presence of surrogates? There are a few available technologies, such as service discovery in Bluetooth, and one that is most suitable for an application should be selected.
- How does one establish an appropriate level of trust with surrogates? Because surrogates are placed in public spaces, it might be relatively easy for malicious parties to tamper with them to retrieve data about the user or the data that has been retrieved or processed for the user.
- How does the system decide which computation to offload to a surrogate? There are numerous load-balancing algorithms and one that is appropriate for an application has to be chosen.
- How much advance notice does a surrogate need to act as an effective staging server with minimal delay?
- How is scalability addressed to avoid overload during peak periods?
- What support is needed to make the use of surrogates seamless and poses minimal intrusion to the user?

Once again, the issues that have to be addressed are nontrivial. At the moment, available solutions are simplistic and might not scale in a complex environment. Much work and research have to be carried out to support adaptability efficiently and effectively.

8.3   Power Management

Because mobile devices operate on battery power, it is imperative that the battery lifetime be extended for as long as possible. Flinn and Satyanarayanan (2000) observed that as more sophisticated applications and services are made available to users, there is an increasing pressure for devices with more powerful processors. Sophisticated capabilities place severe restrictions on battery capacity. At the same time, there is the pressure of making mobile devices smaller, lighter, and more compact. These two contradicting requirements are very hard to reconcile.

Power conservation strategies can be divided into two approaches: hardware and software. The hardware approach to reduce power consumption involves powering down power-hungry components such as the hard disk, the display, and the CPU. The software approach involves offloading heavy-duty computing to a fixed host and designing software that is able to adapt its power consumption.

8.3.1 Power Conservation Strategies

There are three components that consume a significant amount of power: CPU, hard disk, and display. The CPU consumes 16–35% of total power. The hard disk consumes 8–22% of total power. The display accounts for 44–60% of total system power. Powering down one of these components can reduce power consumption between 11–60%.

Power conservation strategies for hard disks involve techniques for spinning down the disk during an idle period. The concerns that have to be addressed when spinning down the hard disk are:

■ Establishing an appropriate threshold for the idle period so that the disk can be spun down safely.

■ Keeping access latency to a minimum because once a disk is spun down, a delay is introduced when spinning up the disk, thus delaying access to data.

■ Spinning down the disk to reduce power consumption should not adversely affect the disk life span.

■ Spinning down the disk should not result in higher power consumption compared to when it is not spun down.

An adaptive disk spin-down policy (Douglis, Krishnan, and Bershad 1995) balances between reducing power consumption and reducing access latency. The algorithms distinguish between undesirable and acceptable spin-up delays. Undesirable delays are referred to as bumps and the timeout value was varied based on users' tolerance of bumps. The adaptive policy aims to reduce the number of bumps without adversely affecting energy consumption compared to a fixed-threshold policy or reduce energy consumption without adversely affecting the number of bumps.

Marsh and Zenel (1994) used three different strategies to power down the CPU. The strategies, in increasing order of complexity, are:

■ Halt: A technique for reducing CPU power by executing the halt instruction. Once halted, most of the transistors do not change state, thereby reducing power significantly. The halt instruction is simple and consists of only 6–10 assembly instructions. The CPU still runs at full speed because the clock rate is not altered. When halted, no instructions are fetched and nothing happens until an interrupt, which may be generated by the RTC or any peripheral devices, is sent. Halt reduces total system power consumption by about 22%.

■ Clock: Relies on hardware and its power management chip. When the idle thread finds no other threads in the ready queue, it reduces the clock speed. The ready queue is continually inspected at the lower speed until a new thread appears. The clock is then reset to full speed and control is transferred to the new thread. The clock speed is also increased if an asynchronous software trap is received. This approach reduces power consumption by an additional 8–11%.

- Clock/Halt: First reduces the clock speed and then calls the halt instruction. Unlike the first approach, some logic may change state when halted. This approach reduces the CPU's power consumption while waiting for an interrupt. The idle thread checks the ready queue and reduces the clock speed and calls the halt instruction if there is no ready thread. When an interruption is received, the chaining code turns the clock back to full speed. When the interruption completes, control returns to the instruction following the halt instruction. This technique performs best, reducing power consumption by 35%.

Because the CPU is one of the most power-hungry components, reducing computation on a mobile device helps to reduce power consumption. One approach to reduce power consumption by the CPU is by off-loading computation from a mobile device to a host on a fixed network (Othman and Hailes 1998; Rudenko et al. 1998). In this strategy, an adaptive load-sharing algorithm is used to select jobs for remote execution. Offloading computation to fixed hosts was found to extend battery lifetime by 20% (30–60 min). The offloading of computation is transparent to the user.

Remote execution is also a strategy adopted by Spectra to reduce energy consumption. Spectra is a component in Aura that is responsible for remote execution (Flinn et al. 2001). It monitors resources, such as battery, file cache state, and available bandwidth, and dynamically balances energy use and performance concerns. It is designed to be self-tuning so that applications do not have to specify their intended resource usage. When deciding where an execution should be done, Spectra considers the goal of minimizing application latency while maximizing battery lifetime.

Spectra considers the following factors when deciding where to execute an application: performance, energy conservation, and quality. A resource-poor mobile device may only be able to provide a low fidelity version of data, whereas a stationary machine may be able to deliver a higher fidelity version. It monitors environmental conditions and adjusts the relative importance of each goal. For example, if the device has just been recharged, providing the best performance is a high priority but as energy drains, Spectra trades off lower performance to prolong battery lifetime.

Spectra consists of four resource monitors that predict how much of a resource an operation requires for both the local machine and remote servers; for example, it predicts the bandwidth required and the delay between the client and each server. The prediction is gathered in a resource snapshot that provides a consistent view of resource availability for that operation. The resource monitors observe application behavior and measure resource usage while an operation is executing. When the operation is completed, the observed values (e.g., input parameters, bandwidth usage fidelity) are logged and are used to predict future resource usage. The accuracy of a prediction increases over time as the operation is repeatedly executed. There are four resource monitors:

1. CPU monitor: Predicts CPU availability using a smoothed estimate of recent CPU load. While an operation is executing, the CPU monitor measures CPU cycles consumed on local and remote machines.
2. Network monitor: Predicts available bandwidth and round-trip times of remote machines.
3. Battery monitor: Predicts energy levels by querying the amount of charge left in a smart battery periodically. A smart battery contains chips that report information such as battery level and power drains.
4. Cache state monitor: Because a cache miss results in data access that consumes time and energy, a cache state monitor estimates the cost of fetching a file that is not available in the mobile device's cache. During an execution, it monitors file access and upon completion, it logs the name and size of each file accessed. The prediction scheme assumes that the likelihood of a file access during an operation is similar to the percentage of times it was accessed during recent operations of similar type and input parameters. The access likelihood is maintained as a weighted average and is adjusted by the monitor according to changes in application behavior over time.

Spectra's decision engine chooses a location and fidelity for each operation. It evaluates alternatives by their impact on user metrics that measure the end user's perceived performance or quality. An alternative is evaluated based on a context-independent value of each metric. Each value is assigned a weight that reflects the current desirability of the metric to the user. Finally, Spectra calculates the product of the weighted metric to compute a single value for evaluating the alternative. It may use many resource predictions to calculate a metric's context-independent value. For example, execution latency is the sum of the predicted latencies of fetching an uncached item, network transmissions, and processing on local and remote machines.

Odyssey uses a predictive approach that is slightly different from Spectra (Flinn and Satyanarayanan 2000). One of the important characteristics of mobile software is the ability to learn a user's habits and preferences so that this information can be used by the software to anticipate a user's needs. This ability is important not only to deliver just-in-time information but can also be used to conserve power.

Odyssey monitors a user's habits and preferences and uses that information to predict how much power is required to execute the user's tasks. When there is a significant mismatch between predicted demand and available energy, it notifies the application to adapt. This approach extends battery lifetime by up to 30%. The exponential smoothing function used is:

$$\text{new} = (1 - \alpha)(\text{this sample}) + \alpha.\text{old},$$

where α is a parameter that determines the relative weights of current and past power usage. It is varied as energy drains to trade off between agility and stability. As α decreases, the application is biased toward agility.

Future power usage demand is estimated periodically. The estimate is multiplied by the time remaining to complete the task to estimate future energy demand. If predicted demand exceeds residual energy, Odyssey issues an upcall so that the application can adapt to reduce energy consumption. Conversely, if residual energy significantly exceeds predicted demand, applications are notified to increase data fidelity. When multiple applications run concurrently, Odyssey decides which ones to notify based on a user-specified priority scheme.

Odyssey defines a hysteresis in its adaptation strategy. Supply must exceed demand by a defined amount to trigger fidelity improvement. The hysteresis value consists of two components: a variable (5% of residual energy) and a constant (1% of the initial energy available). The variable component is biased toward stability when there is ample residual energy and toward agility when it is scarce. The constant component is biased against fidelity improvements when residual energy is low. Fidelity improvement is restricted to once every 15 s to avoid excessive adaptation. Odyssey's adaptation strategy varies for different components.

For video player applications with hardware-only power management, energy consumption reduces by 9–10% mainly because the hard disk remains in standby mode while the video is fetched and played. A lossy compression is used to reduce video fidelity, resulting in 16–17% less energy power consumption than using hardware-only power management. Halving the height and width of the display window further reduces energy consumption by 19–20% than using hardware-only power management. When reducing video fidelity and the size of window display is combined, energy consumption is reduced by 35%.

The speech recognizer application consists of a front end that generates a waveform based on the word uttered by the user. The waveform is submitted to a local or remote speech recognition system for processing. A local recognition incurs a processing cost but saves on transmission cost, whereas a remote recognition incurs energy cost for transmission but can exploit the CPU, memory, and energy resources of the remote server. A hybrid operation is also possible where the first phase of recognition is performed locally and a compact intermediate representation is then sent to a remote server for the remaining recognition process. In this case, fidelity is lowered by using a reduced vocabulary and a less complex acoustic model resulting in reduced memory footprint and processing required for recognition. The down side is it degrades the recognition quality. Although hardware-only power management reduces energy consumption by 33–34%, lowering fidelity results in a reduction of 25–46%. Remote recognition at reduced fidelity results in a saving of 42–65%. A hybrid operation at reduced fidelity results in a saving of 53–70%. Combining hardware-only power management with hybrid, low-fidelity recognition results in 69–80% less energy consumption. The optimal strategy to adopt depends on resource availability and the user's tolerance for low-fidelity recognition.

An adaptive map viewer named Anvil fetches maps from a remote server and displays them to the client. Fidelity is lowered by filtering that eliminates fine detail and less important features and cropping that restricts data to a geographic subset of the original map. Hardware-only power management reduces energy consumption by 9–19%. When combined with filtering, a reduction of 6–51% is achieved, depending on the level of filtering applied. Combining filtering and cropping results in an energy reduction of 36–66%.

Zone backlighting is an experimental power management approach where the screen is divided into grids; each grid representing a zone. The energy usage of each grid is controlled independently. When residual energy exceeds demand, a large high-fidelity image spans several zones. As the battery drains, applications generate smaller, lower-fidelity images that span fewer zones. This approach saves power consumption by 7–29%.

The various power conservation techniques discussed above may be combined to prolong battery lifetime significantly. Even though power conservation is undeniably a critical element of mobile computing, is there an alternative to returning home to recharge a flat battery? Yes, there is. The technique is known as energy scavenging.

8.3.2 Energy Scavenging

Energy scavenging, also known as energy harvesting, is a technique that collects energy from its surroundings to power systems and it may involve storing energy when it is not required in batteries, capacitors, or springs. The most familiar example of energy scavenging is solar cells used to power calculators. Energy scavenging is especially useful for applications that require small amounts of continuous power or that have short periods of high-power use (Want et al. 2005). Energy may be scavenged from a number of sources (Paradiso and Starner 2005):

■ Background radio signals: Electronic systems may harvest energy from ambient-radiation sources. A drawback of this approach is it requires either a large collection area or close proximity to the radiating source. The amount of energy harvested is on the order of tens of microwatts.
■ Ambient lighting: The energy conversion efficiency of a solar cell module is less than 20%. Utilizing sunlight is not a problem in tropical regions, but the lack of strong, consistent sunlight in other regions may pose a constraint. The amount of energy harvested from sunlight varies from a few kilowatts for a solar home system to up to 2 W for a cell phone that harvests power from direct sunlight.
■ Thermoelectric conversion: This involves harvesting energy from objects or environments at different temperatures via heat transfer. An example of

thermoelectric conversion is the Seiko Thermic wristwatch that uses ten thermo-electric modules to generate energy (a few microwatts) from the small thermal gradient provided by body heat over ambient temperature.

■ Vibrational excitation: Vibrations vary widely in frequency and amplitude. Among the sources of vibration are vibrations on floors and walls caused by nearby machinery, from jet engine housings, and automobile chassis. Energy is harvested by exploiting the oscillations of a proof mass resonantly tuned to the environment's dominant mechanical frequency. An example of harvesting energy from vibrations is a self-winding wristwatch that contains a rotary proof mass mounted off-center on a spindle. As the wearer moves, the mass spins and winds the mechanism.

■ Power from human input: Actions such as cranking, shaking, squeezing, spinning, pushing, pumping, and pulling can be exploited to power up a device. A piezoelectric element with a resonantly matched transformer and conditioning electronics generate approximately 1 mJ at 3 V per 15-N push when struck by a button. The power generated is enough to run a digital encoder and a radio that can transmit over 15 m.

■ Ambulatory power generation: A human body generates about 0.1 to 1.5 kW^2 when it moves from resting to a fast sprint. A person exerts up to 130% of their weight across their shoes at heel strike and toe off. Standard jogging sneakers compress by up to 1 cm during a normal walk, meaning that for a 70-kg person, about 7 W of power is available at a 1-Hz stride from heel strike alone. Walking produces an average of 250 to 700 mW.

The amount of power that can be scavenged is limited by the amount of available raw energy and the surface area of net mass of a device. The available power is sufficient for devices that require low power as available power is usually around milli- or microwatts. Even though currently the amount of power that can be scavenged is very little, hopefully ongoing research in this area will offer better results in the future.

8.4 Interface Design

The most visible constraint of mobile devices is the display size. Even though higher data rates are now available to support multimedia applications and faster data downloads, the amount of data that can be displayed will still be constrained by the small display size.

The small size of mobile devices imposes a constraint in I/O method. Some mobile devices support extensive graphic capabilities, and others may provide limited display resolutions. Some are equipped with enhanced I/O devices, such as a portable keyboard, and others are constrained by limited input, for example, a stylus pen. Mobile phones are not equipped with a keyboard and each button represents three to four characters, making it slower and inconvenient to input text. Consequently, UI design not only needs to take these constraints into consideration, it also needs to consider how the mobile devices will be used to ensure that appropriate interfaces are provided to support users' tasks.

There are a number of approaches to make efficient use of limited screen size, such as minimizing the use of buttons, menus, and scroll bars, all of which consume a considerable amount of space (Grasso 2004). These objects may be replaced with objects that can be used for both navigation and providing context, for example, focus-and-context visualization that divides the screen into a set of objects. A central object provides details, and other objects provide thumbnail overviews used for navigation.

Currently, the methods available for text input are still unsatisfactory. Given these restrictions, the following must be considered during the design and development of mobile applications:

- Input should be restricted to a minimum: Avoid alphanumeric input as far as possible. If it is unavoidable, defer it to the latest possible point in time. Drop any functionalities that require a lot of alphanumeric input.
- Focus on very important information: Even though there is a significant increase in bandwidth with 3G technology, the amount of useful information that can be displayed on a mobile device is still limited. Information should be customized to the needs of the user.

Even though text input may be unavoidable in certain applications (e.g., password input for mobile banking), it may be inappropriate for other types of applications as illustrated in the case study discussed in Section 8.7.1.

Desktop computers allow a GUI that often relies on rich visual feedback and requires a user's undivided attention—a luxury on mobile devices considering that a mobile user is often occupied with multiple tasks (e.g., driving while checking traffic or weather report). Consequently, an interface for mobile applications should be one that is not distracting and not demanding of users' attention.

Users may use different platforms to perform a tasks. For example, a user may prepare and share a presentation using a desktop computer. On the way to meet a client, the user may review the presentation and make a few changes using a pocket PC based on comments received from a colleague. The updated presentation is then downloaded to a laptop just before the presentation to the clients. This scenario illustrates that UIs for mobile applications need to be sensitive to (Eisenstein et al. 2002):

- Platform: Adapts to the characteristics of the device, such as screen surface, color depth, and screen resolution. Interactive dialogs must consider available bandwidth and adapt accordingly.

■ Interaction: Provides a mechanism that remembers previously used interaction techniques, for example, window size and location.

■ User: Adapts to user skill level, system and task experience, and preferences.

In the following sections, two studies are used to illustrate various aspects of mobile interface design.

8.4.1 Context-Aware UI Modeling

The Map ANNotation Assistant (MANNA) is a multimedia application that is used by geologists, engineers, and military personnel to create annotated maps of geographical areas (Eisenstein et al. 2002). Due to the nature of the application and its intended use, the application has to run on several platforms and can be utilized over the Internet. Hence, the UIs must be tailored for use on the different platforms.

Eisenstein et al (2002) describe the following possible usage scenario. A geologist has been dispatched to an area that had recently been affected by an earthquake. Before leaving, the geologist uses a desktop computer to download maps and reports of the area. The geologist also downloads the relevant information to a laptop before heading for the site. When the geologist uses the laptop on the airplane, commands that rely on a network connection are disabled because the laptop is not connected to the Internet. When the geologist examines a video of the site, the UI switches to a black-and-white display and reduces the rate of frames per second to conserve battery power. At the airport, the geologist receives a message from MANNA via mobile phone with alerts to examine a location. The map of the region is not displayed due to display limitation, but driving instructions to get to the location are sent along with the geologist's current GPS location. At the site, the geologist prepares a report using a palmtop. Because the palmtop relies on a touch pen for interaction, the UI avoids double-click and right-click interactions.

There are three model components for designing a UI for mobile computing applications:

1. Platform model: Describes the various computer systems that run a UI, including constraints placed by the platform on the UI. This model may be static or dynamic. A static model allows the UIs to be exploited at design time and a set of UIs can be generated for each platform. On the other hand, a dynamic model is exploited at run time and is sensitive to changing conditions; for example, it responds to changes in available bandwidth by adapting the interface accordingly.

2. Presentation model: Describes the visual appearance of the UIs, such as the hierarchy of windows, widgets, stylistic choices, and the placement of widgets.

3. Task model: Presents a structured representation of tasks that the user may perform. It is hierarchically divided into subtasks and may include information on goals, pre- and postconditions, and whether a task is optional, may be repeated, or enables another subtask.

The adaptation of UIs may occur at design time in the form of automated design support or at run time via a dynamic UI. Two techniques are used to design adaptive UIs such as the ones given in the scenario above: abstract interaction objects (AIOs) and concrete interaction objects (CIOs). In AIO, an interaction object is an element that allows users to visualize or manipulate information or perform an interactive task. The objects are the atomic building blocks of a UI. An AIO is executable on any platform and is implementation-independent and, thus, portable. AIOs allow the implementation of the same UI on multiple platforms. Unlike an AIO, a CIO is an interactor, such as a dialog box or a button, that is executable on a platform without additional processing. An AIO is a parent of several CIOs. A CIO inherits the property of its parent AIO and is mapped on to the platform it supports. A designer need only be concerned with the AIOs as the appropriate CIO is rendered automatically depending on the platform.

Mobile devices come in different screen sizes and resolutions. There are three parameters that need to be considered when determining the amount of display size required by a UI: size of individual interactors, layout of interactors within a window, and the allocation of interactors among several windows. There are two ways to accommodate screen resolution constraints by reducing the size of interactors:

1. Shrink the interactors while observing usability constraint related to the AIO type; for example, reduce the length of an edit box to a minimum while the height is maintained at least at the smallest legible font size.
2. If the size of an interactor cannot be reduced, the interactor is replaced with a smaller alternative; for example, a Boolean check box requires less screen space than a pair of radio buttons.

The layout of interactors within a window and the allocation of interactors between windows are grouped under a presentation structure. The appropriate presentation structure is selected based on the constraints of available screen space. There are two solutions. The first solution involves creating mappings between each platform and an appropriate presentation structure. This approach is simple but the drawback is that it is static; for example, if the available screen solution increases at run time, the application is unable to take advantage of this change. The second solution involves building a mediator agent that dynamically selects the appropriate presentation model for each device.

Although it is possible for a designer to specify a set of presentation structures, it is better if the correct structure is generated by the system within the constraints of AIOs. For this purpose, there are two abstractions:

1. A logical window (LW) is a grouping of AIO, such as a window area or a dialog box. All LWs are constrained by the screen space.
2. A presentation unit (PU) is a complete presentation environment required to execute an interactive task. A PU may be decomposed into one or more LWs that may be displayed on the screen simultaneously, alternately, or a combination of the two. Each PU consists of at least one window called the main window that enables navigation to other windows.

It is often assumed that the same functionalities should be provided on every type of device to support the same set of tasks. However, this is unnecessary as a user often performs only a subset of tasks on one type of device. For example, referring to the scenario given earlier, a desktop is used to obtain a comprehensive map of the area, a mobile phone is used to obtain positioning information and driving directions, and a palmtop is used to prepare a report on site. Each device is used to perform a subset of tasks. Using a task model (Figure 8.1), it is possible for the system designer to take advantage of the knowledge about what tasks are performed on which device to create mappings between platforms and tasks.

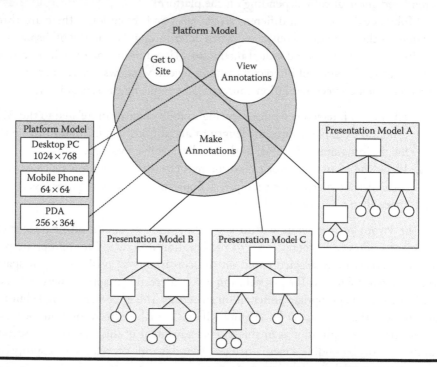

Figure 8.1 Task, platform, and presentation models. (From Eisenstein, J., J. Vanderdonckt, and A. Puerta. 2002. Adapting to mobile contexts with user-interface modeling. *Proceeding of the 3rd IEEE Workshop on Mobile Computing Systems and Applications.* Used with permission.)

The presentation model can be optimized to perform a subset of tasks. Tasks that have been identified as important should be represented by AIOs that are easily accessible. For example, clicking on a map while using a palmtop allows the user to enter a note whereas doing the same while using a desktop displays a rich set of geographical and meteorological information of the selected region in addition to displaying a list of previously entered notes.

8.4.2 Tactile Interface

An alternative to using keypad and stylus pen as a means of input is to use the sense of touch. This can be accomplished by embedding a PDA with a tactile actuator named TouchEngine (Poupyrev et al. 2002). Because mobile devices are often kept close to the user's body, the sense of touch provides a means of interaction that is nonobtrusive. Tactile interface is a good channel of interaction for mobile devices for the following reasons:

- It is fast: The sense of touch is five times faster than vision and humans can detect two consecutive tactile stimuli in about 5 ms.
- It needs little conscious control: Because large areas of the sensory cortex are devoted to processing stimuli from the skin, complex motor operations produce little cognitive load (a large cognitive load demands more attention from the user, thus, distracting him from other tasks) and can be performed in parallel with other activities.
- It allows for information encoding: Touch allows us to ascertain object properties more accurately, even where vision fails, for example, texture variation. With proper encoding, messages can be transmitted through the skin. Different vibration patterns can be used to convey different messages (e.g., task reminder, incoming email).
- It produces strong emotional responses: We can often recall how a touch feels (e.g., stroking a child's hair). Touch has a very strong impact and if aroused in unusual patterns can be attention-demanding.

Haptics is a term used to refer to tactile feedback. The development of a haptic display involves:

- Choosing or developing a haptic actuator, which is a transducer that converts electrical signals into mechanical motion. Electrical motors are often used as actuators in haptic research.
- Designing a tactile display that converts mechanical motion produced by the actuator into force communicated to the user. The same actuator can produce various haptic displays with very different interactive properties.

■ Developing control hardware and software. Effective operation of a haptic display requires a good understanding of its properties.

There are two arguments for ambient or peripheral awareness displays. First, they allow users to perceive and process incoming information with little conscious effort and without interrupting current activities. Second, they provide unobtrusive notification techniques to grab the user's attention when the interruption is needed.

Touch is the perfect ambient information channel because it is more powerful than vision or sound. We already rely heavily on our tactile sense to receive information from the background; for instance, we feel the ground with our feet and unconsciously adjust how we walk when we step from the pavement to the grassy lawn. We may not hear our name called, but we react when tapped on the shoulder. Most of these are done with little focused attention. Ambient tactile interfaces may be particularly useful in mobile applications because tactile displays that are close to the body allow information to be received without interrupting the current activity.

Even though mobile devices utilize vibrotactility for notification, the current tactile displays convey only one bit of information. For example, when a silent mode is activated, a hand phone vibrates to indicate an incoming call, but it does not indicate the urgency of the call or who the caller is. Consequently, the user has to suspend his current task to interact with the device and determine who the caller is and the urgency of the call. An approach to overcome this problem is by utilizing intelligent context-aware and ambient notification mechanisms. It has been shown that humans can accurately distinguish different vibrotactile patterns. TouchEngine provides six tactile pulse sequences with different rhythms, intensities, and gradients of vibration.

Consider the task of downloading a file to a desktop computer. You can periodically glance at the progress bar to monitor the download as it progresses and determine if the download has been completed. This visual monitoring process is easy to do on a desktop computer. However, if you are downloading a file to your pocket PC, you might not be able to visually monitor the download progress. It would be more convenient if you could initiate the file download and then put away the device in your pocket while you carry out other tasks, such as buying a train ticket and walking to the platform to wait for the train. An alternative to the visual progress bar is a tactile progress bar that vibrates when the download is completed. The tactile progress bar provides a short tactile impulse by clicking periodically and the intensity of click increases as the download nears completion. Once the download is completed, you can decide if you want to retrieve the device to peruse the downloaded file or wait until you get a seat in the train to do so. Either way, the task of downloading the file does not demand your full attention and does not distract you from other tasks you have to perform.

Tactile and tilting interfaces may be combined to provide a more effective means of interaction. The device with tilting interface is equipped with tilt or pressure sensors, and users interact with them by physically manipulating the device; for example, a user tilts a mobile device back and forth to scroll through a page. An advantage of the tilting interface is it does not require a two-handed input, thus, freeing one hand for other physical tasks. A drawback of the tilting interface is users tend to overshoot the target destination; for example, upon reaching the intended paragraph in a text, users could not return the device to its neutral state in time to stop the scrolling. As a result, users tend to tilt in the opposite direction in an attempt to stop the scrolling. Another drawback is it requires constant visual attention.

An example of how tilting and tactile interfaces can be combined is in scrolling a two-dimensional (2D) map. The map scrolls in the direction of the tilt. Each time the image moves on the screen, a simple scratching tactile pattern is played so that the user can feel the map shifting within the device. The tactile pattern changes when the image reaches its boundary. Another example is the use of a tactile tap each time one line of text is scrolled up or down. The speed of the tapping becomes slower or faster accordingly as the user changes the scrolling speed. When the device is in a neutral position, the interface produces a short buzz so that the user is aware that the device is in its neutral position, thus, reducing overshooting when the user wants to stop scrolling.

The haptic interface is also useful in clinical applications on handheld computers (Grasso 2004). A patient's medical chart can be picked up, carried, held in one hand, and annotated with a pen. The same can also be done with a PDA or palmtop. However, there are certain things that cannot be done with a handheld computer, such as creasing a page to mark it or flipping through pages. When designing a mobile application, it is imperative that the application does not impose too much change in the way the users have got used to performing their tasks. To achieve this, the handheld device is incorporated with a haptic display with the properties of a patient's chart. Instead of turning the pages by scrolling, the physician does so by brushing a finger across the top corner of the display—similar to turning a page in a book. The handheld device is equipped with tilt and pressure sensors that measure the angle of the device and finger pressure along to edge that give a similar effect as fanning through a book.

8.5 Heterogeneity of Devices and Environments

User devices vary in terms of screen size, color display, effective bandwidth, processing power, and the ability to handle specific data encodings. These variations make it difficult for mobile application developers to make any assumptions regarding the capability of the devices that use their applications. One approach to handle the variations is by using on-demand distillation to increase the QoS to the client and reduces latency. Instead of performing the distillation at the client (device) or the server, it is performed in the network infrastructure. Distillation is defined as highly lossy, data-type specific compression that preserves most of the semantic content of data object while adhering to a particular set of constraints (Fox et al. 1996). The on-demand distillation considers the following variations:

- Network variation: In terms of bandwidth, latency, and error behavior. On-demand distillation focuses on reducing bandwidth requirements at the application level.
- Hardware variation: Such as screen size and resolution, color and grayscale bit depth, memory, and processing power.
- Software variation: In terms of application-level data encoding that a device can handle (e.g., JPEG, nonstandard HTML extensions) and protocol extensions such as IP multicast support.

The distillation is data-type specific. Fox et al. (1996) considered the distillation of video, image, or text. Lossy image compression involves discarding color information, high-frequency components, or pixel resolution. In addition, lossy video compression includes frame rate reduction. Lossy text compression involves discarding some formatting information but preserving the actual prose. The distillation tries to preserve information that has the highest semantic value. Knowledge of the data type also allows the reordering of traffic to minimize perceived latency. For example, if a user is reading a news update, text is given higher priority than images so that information of the highest semantic value is delivered to the user first.

The purpose of a distilled object is to allow the user to evaluate the value of the downloaded object and then request a higher-quality presentation of a specific section of the object later, if the user wishes to have it. For example, the user may zoom in to a particular section of an image and request a high-quality presentation of that section instead of the entire image, thus, putting a lower demand on bandwidth requirement. The distillation is performed by a proxy and this approach offers a number of advantages:

- Instead of storing a few versions of an object to accommodate device capability variations, servers store only high-quality content.
- Servers do not bear the cost of performing the distillation.
- No changes are required on legacy servers.
- Simple and inexpensive clients can rely on the proxy to optimize content from servers designed for higher-end clients.
- Clients communicate only with the proxy instead of possibly multiple servers.

■ A service provider may offer distillation and refinement as a value-added service, resulting in an economic model favorable to the service provider, clients, and servers.

The proxy uses the distillers to optimize the QoS for the clients in real time. The network connection monitor (NCM) monitors end-to-end bandwidth and connectivity to the proxy's clients and notifies the proxy of any changes that may affect the proxy's transcoding decisions. Client applications are linked with an APS library that provides a standard API for communicating with the proxy. Proxy-unaware applications communicate with the proxy via a client-side agent. The NCM determines a client's network connection characteristics by tracking the value of effective bandwidth, round trip latency, and the probability of packet error. Alternatively, the client may notify the proxy of its expected bandwidth.

A client does not communicate directly with the server; it only communicates with the proxy. The proxy retrieves content from Internet servers on the client's behalf, determines the high-level types of the various components (e.g., images, text runs), and determines which distillation engines to employ. When the proxy calls a distiller, it passes information such as the hardware characteristics of the client, acceptable encodings, and available network bandwidth. This information is used by the distiller to optimize the object representation; for example, the distiller optimizes an image according to the display size. A user may specify a maximum acceptable end-to-end latency for the delivery of an image to reduce end-to-end latency.

8.6 Seamless Mobility over Heterogeneous Wireless Networks

The explosive growth of wireless communications has resulted in several independent and autonomous networks. As mobile users move, they encounter different types of networks offering different levels of services. For example, they may move from a WLAN environment that offers tens of megabits per second to a cellular network that offers tens to hundreds of kilobits per second. The type of services available and the QoS offered also vary. To provide seamless mobility, these variations should be masked from the users so that they receive integrated services.

One of the goals of a mobile computing environment is to mask the heterogeneity of the underlying technologies by presenting a uniform environment to the users. A mobile device with multiple network interfaces should be able to move between different proprietary networks seamlessly without disturbing the applications. The applications are equipped with system support that are capable of adapting to dynamic QoS variations gracefully. Ideally, the services that are available to the user in a WLAN should also be available via cellular service albeit at a reduced QoS level. To support seamless mobility, adaptive computing is required.

A graceful reaction to sudden QoS changes is possible if the environment and applications collaboratively adapt to the dynamic operating conditions.

To provide support for seamless mobility, the following factors have to be considered:

■ A wireless network is often autonomously owned and operated. It is unlikely that its BSs will be accessible for software changes to support seamless mobility. Whatever solutions must operate using off-the-shelf commercial networks.

■ Wireless networks may vary in terms of bandwidth, latency, channel error, range, access protocol, failure probability, connection cost, pricing structure, and security.

■ Mobile devices have varying capabilities. Among factors to consider are computing power, memory, network interfaces, disk space, compression techniques, battery power, and operating system.

■ A mobile device will encounter a variable QoS depending on its current connections due to the dynamic nature of the environment.

As a user moves, connections may be set up or torn down, and this may also involve QoS changes. When moving within the same network, a user may move from a noncongested cell to a congested cell. There are three application-level approaches that can be used to deal with QoS variation:

1. Applications support a seamless mobile computing environment without notification of QoS change. This approach does not work well when QoS changes by orders of magnitude, for example, from a WLAN environment into a WMAN environment.
2. Applications are allowed to (re)negotiate QoS with the network. If the network could not honor the negotiated QoS, the application is alerted.
3. Applications list acceptable QoS classes, which is the procedure to execute if a QoS class is granted and the exception handling policy if the network violates the QoS agreement. The mobile computing environment handles the QoS (re)negotiation and reaction to notification.

There are three approaches to support dynamic QoS (re)negotiation:

1. The mobile computing environment provides the measured QoS in a format that can be retrieved by the applications. Applications poll the environment periodically to retrieve the current value and adapt accordingly.
2. As an addition to the first approach, an application registers with the environment its acceptable QoS bounds. The application is notified if the bounds are violated and adapts accordingly.
3. The application program is split into ranges separated by QoS system calls. In each call, the application specifies a sequence of acceptable QoS classes,

the procedure to execute if each class is satisfied, and the exception handling if there is a change in the QoS level. In case there is a change in the QoS level, notification is handled by the exception handling procedure, which may either block, abort, rollback, or continue.

A mobile device may have access to multiple wireless networks at a time; for example, a PDA may be able to access the local WLAN service and at the same time is able to access services provided by the user's 3G network operator. Which network to choose for data transport depends on the application traffic characteristics, wireless network characteristics, transmission priority, and cost. Migration between networks occurs for a number of reasons, such as user mobility, network failure or partition, or application-dependent tradeoffs. Seamless mobility requires that the process of switching between networks is done transparently. There are four levels of mobility:

1. Handoff within an organization and a network: A mobile user moves between two cells of the same service provider within a network.
2. Handoff between organizations but within the same network: A user switches between service providers on the same network, for example, switches carriers while using the same cellular phone line.
3. Migration within the organization but between networks: A mobile user switches between networks of the same service provider, for example, between a cellular phone and a cellular digital packet data service of the same carrier.
4. Migration between organizations and networks: A mobile user switches both networks and service provider, for example, switches between locally owned WLAN to a GSM service.

To support seamless mobility, the following constraints must be observed. A mobile user may migrate between different networks that are owned and operated autonomously. The networks may use different protocol stacks, especially on the lower layers. Service stations on different networks may not be able to communicate with each other directly for mobility support. Finally, it is likely that each network has its own unique address for the mobile device.

8.7 Other Issues in Mobile Application Design

When developing an application, whether for desktop computers or mobile devices, it is imperative for the system designer to understand not only the requirements of the application, but also how the target users will use the applications. What makes designing mobile applications more challenging is the fact that a number of assumptions about developing desktop applications are not valid in a mobile environment.

Designing and developing mobile applications are significantly different from designing applications for desktops due to various constraints of mobile devices and the mobile environment. Limitations on the device display, processing capability, and bandwidth have to be taken into consideration. Moreover, the designer has to keep in mind that mobile devices are not used in the same way as desktop computers are used. For example, while a salesperson might have no problem using a PDA with a stylus pen for input when updating a database, a service technician at a car manufacturing plant might find it inconvenient to use a mobile device that requires both hands for data input. The technician might prefer a means of input that is hands-free and does not demand much attention. When designing a mobile system, the physical configuration should be determined by its operation. This is just one example of how a thorough understanding of how users carry out their tasks and how they interact with one another in a social and organizational setting is critical in designing a good and effective mobile application.

In this section, you will learn about various constraints and considerations that must be kept in mind when designing mobile applications. You will see that the factors to consider may vary greatly depending on the applications and their operating environments.

8.7.1 Mobile Service Design for Service Technicians in Industrial Settings

An ethnographic field study was carried out on Volvo cars and trucks to determine what service and maintenance personnel did, whom they talked to during their shift, and their work practices (Fallman 2002). The information obtained was later used to develop a mobile information system prototype that can be used by the personnel without turning it into a work procedure. Table 8.1 summarizes the observations and design incentives of this study.

The prototype developed is a PDA-based, arm-worn, gesture-driven, and context-aware embodied system that supports useful and interconnected activities for service and maintenance work in industrial assembly settings. The device is attached to a user's arm, allowing the user to interact directly with objects in the physical environment. Users use their hands to point directly to a physical object to initiate interaction. The prototype is developed using a customized Compaq iPaq H3660 with a 206-MHz Intel Strong ARM 32-bit processor, a color reflective thin film transistor liquid crystal display, 64-MB memory, and a RS323 serial communication port. The operating system is Microsoft Pocket PC 2002. The prototype is also equipped with a RF ID tag (RFID) reader and a 2G accelerometer sensor, which are put in a physical glovelike configuration worn on a user's arm.

The context-aware functions provided are component history and virtual notice board functions. The history function allows the technician to access all previous maintenance and service carried out on a component. The virtual notice boards allow technicians to leave voice notes, for example, to explain a problem about a component and what action has been taken. This function reduces the level of

Table 8.1 Summary of observation and design incentive of the case study.

Observation	Design Incentive
Service technicians did not repair components that broke down. Because the components were sent away for repair, there was no need for mobile access to manuals or extensive help on repair work. However, the technicians would benefit with help on how to remove, install, and replace components. Moreover, a new employee often had difficulty with procedures that suggested many different types of installations.	Equip service technicians with digital access to blueprints because manuals were rarely used.
When a component broke down, the technician often used the DECT phone to call the manufacturer of the component to speak to a person with the appropriate expertise to deal with the problem. An alternative was to use a graphical blueprint instead of a manual for troubleshooting.	Provide quick and easy access to manufacturers of components.
Service technicians had a problem determining the availability of spare parts. If a component had to be replaced, the technician had to either go to the office to use a desktop computer to determine whether the spare part was available or go to the stock room to find the spare part.	Provide quick access to spare part inventory to allow the technicians to decide whether to carry out an on-site repair, to immediately replace a component with a spare part, or to troubleshoot the problem offline. Another benefit of providing this information is that an order can be placed if a spare part is depleted.
Service technicians commented that they would benefit from information about the history of each component; for instance, if a component had been replaced a few times within a week, it indicated that the problem was not with the component but something else. At the time, this information was conveyed using a notice board. This was not an efficient method as sometimes a repair or replacement of a component was not documented immediately and, thus, was unknown to other technicians. Notebooks were scattered throughout the factory but were largely unused because they were not physically close to the assembly cell.	A history of a component is provided in front of the component instead of having to look it up at the office. The system incorporates a virtual noticeboard or notebook so that it can be close to an object (component) that it describes. By making it virtual, a technician can access all notebooks via one mobile device.
Service technicians did not have a complete list of telephone numbers of people they had to collaborate with. When they had to find a number quickly, they often asked the nearest manager (managers were equipped with PalmPilot PDAs) to look up the telephone number.	Service technicians are provided with a list of useful telephone numbers that are divided into a few categories, such as managers, forepersons, manufacturers, and service technicians currently on the shift.

effort required to leave notes (no need to sit down, find the right folder, and write a report) and makes the notes available to other technicians more quickly, hassle-free. The system does not incorporate digital manuals as the service technicians rarely make use of them. They are, however, provided with quick access to manufacturers' contact information so that a manufacturer can be contacted quickly whenever the need arises.

8.7.2 Mobile Commerce

In the past few years, businesses have had to adapt to the use of new technologies to remain competitive. One of the strategies that have been adopted by many companies is e-commerce that has changed the face of the business from physical offices and business facilities to a virtual, click-on-a-Web-site-type of businesses that make it possible for us to spend and invest our hard-earned money from the comfort of our homes.

Recently, another similar transition has emerged and this evolution is marked by an effort to incorporate the Web sites of the businesses into the small screens of mobile phones. Businesses can now present their storefronts not only on Web sites, but also on mobile devices so customers can now spend their money or view similar information without being confined to their desktops. This opens the opportunity for businesses to be conducted anytime, anywhere. This class of application is known as mobile commerce (m-commerce). One classification of m-commerce applications that is becoming more available is the mobile payment service. Examples of services in this category are e-wallet, mobile banking, bill payment, and fund transfer, which can be done via the mobile device. The advances in network and device capabilities have enabled new features of mobile applications that could further increase their usability.

As mentioned earlier, mobile users often perform a few tasks simultaneously. To consider this multitasking behavior, m-commerce interfaces should be designed to support users' limited attention. The interface design has to compensate for the limited visual display of mobile devices. The 7C (context, content, community, customization, communication, connection, commerce) design elements may be adapted when designing an interface for m-commerce (Lee and Benbasat 2003).

1. Context: Links should connect pages seamlessly and efficiently to ease navigation without demanding too much of users' attention. A shallow-structured menu (fewer levels but more choices per level) is preferable because a deep hierarchy imposes a higher cognitive burden by offering more choices over more levels.
2. Content: Refers to what a site presents, such as the offering, appeal, multimedia content and content type. A context-aware functionality may be used

to adapt the product information (offering mix) and promotional messages (appeal mix) offered to users. A proximate selection method is used to emphasize nearby objects and make choices easier. Using audio content instead of text would reduce the amount of display space required.

3. Community: Refers to interactive and noninteractive communications between users. Most people enjoy shopping with friends and like to exchange opinions about products. Interactive communication enables the exchange of opinions and can be realized using, for example, SMS.

4. Customization: Many Web sites allow users to personalize their experience when visiting the sites. Likewise, a user's mobile setting can be used to automatically adapt the mobile interface. This customization reduces the information load by filtering information not of interest to the user, thus, reducing the burden on the limited visual display.

5. Communication: Refers to the dialogue between sites and users. This may be in the form of broadcast, interactive, or a hybrid of the two. Advertising, for instance via SMS, ought to be tailored according to users' context. The advertisement could also take into consideration factors such as time and weather, for example, a special offer on ice cream on a hot afternoon at a nearby cafe. Feedbacks from users may be obtained using multiple-choice answers (often used in SMS quizzes) to reduce the hassle of using the limited keypad. Alternatively, voice and multimedia mail transfer may be used.

6. Connection: Refers to formal linkages between sites that consists of outsourced content, percentage of home site content, and pathways of connections. Pathways to other sites provide users with information needed in dynamic settings. Using location information, an adaptive pathway links only to Web sites of nearby stores to reduce the number of pathways.

7. Commerce: Refers to interfaces related to sales of products and services. A secure payment method that demands minimal attention is required. A one-click checkout can be made available by taking advantage of a user profile that can be configured and later used in subsequent transactions. The user profile consists of information such as user's name, address, and preferred delivery option.

There are a few other factors to consider when developing an m-commerce application. Advertisements should be limited to avoid overwhelming users with information and to avoid the possibility of congestion of the wireless links. If UIs are hard to deal with, users may not see significant value in m-commerce applications. Out-of-context information makes the experience unpleasant to users. Finally, the financial value of m-commerce applications makes security an important aspect to consider and security measures must be thought of carefully when the application is designed. Security should not be something that is added as an afterthought.

8.7.3 Aura: A Platform for Context-Aware Applications

Human attention is a user's ability to attend to primary tasks while ignoring system-generated distractions such as error messages. Project Aura (Garlan et al. 2002) proposed two concepts to minimize distractions to users:

1. Proactivity: A system layer is able to anticipate requests from a higher layer. This is unlike the established approach in current use where each layer only reacts to the layer above it.
2. Self-tuning: Layers adapt by observing the demands made on them and adjusting their performance and resource usage characteristics accordingly.

Garlan et al. (2002) give the following scenarios to demonstrate how proactivity and self-tuning reduces demand on user attention by minimizing distraction.

Scenario 1

Jane is at Gate 23 of an airport, waiting for her flight. She has edited many large documents and would like to use her wireless connection to e-mail them. Unfortunately, connection is very slow because many passengers in the area are surfing the Web. Aura observes that at the current available bandwidth, Jane would not be able to finish sending her documents before her flight departs. Consulting the airport's wireless network bandwidth service and flight schedule service, Aura discovers that wireless bandwidth is excellent at Gate 15, and that there are no departing or arriving flights at nearby gates for half an hour. A dialog box pops up on Jane's screen suggesting that she go to Gate 15 and displays how long it would take her to walk there. It also asks her to prioritize her e-mails so that the critical ones are transmitted first. Jane accepts the suggestion and walks to Gate 15. She watches the news on a nearby TV until Aura informs her that it has almost completed sending her messages so she can start walking back. The last message is transmitted during her walk and she is back at Gate 23 in time for her boarding call.

Scenario 2

Fred is in his office, preparing for a meeting at which he will give a presentation and a software demonstration. The meeting room is a one-minute walk across campus. It is time to leave, but Fred is not quite ready. He grabs his Palm XXII wireless handheld computer and walks out the door. Aura transfers his work from his desktop to his handheld and lets him make his final edits using voice commands during his walk. Aura infers where Fred is going from his calendar and the campus location-tracking service. It downloads the presentation and the demonstration software to the projection computer and warms up the projector. Fred finishes his edits just before he enters the meeting room. As he walks in, Aura transfers his final changes to the projection computer. As the presentation proceeds, Fred is about to display a slide with highly sensitive budget information. Aura senses that this might be a mistake as the room's face detection and recognition capability indicates that there are unfamiliar faces present and warns Fred. Realizing that Aura is right, Fred skips that slide.

In Scenario 1, Aura is proactive in estimating how long it would take for Jane to send her documents. When suggesting that Jane go to Gate 15, it combines knowledge from different system layers. Determining available bandwidth and congestion is a lower layer functionality and determining flight arrival or departure time is an application layer functionality. Aura takes advantage of smart spaces when it determines bandwidth availability at other gates and estimates the time it takes to walk there.

In Scenario 2, the execution state is transferred across diverse platforms: Fred is able to transfer his presentation from his desktop to his Palm to the projection computer. Self-tuning is illustrated through Fred's ability to edit his presentation using speech input instead of a keyboard. Aura is proactive when it infers where Fred is heading based on his calendar and the location-tracking service, warms up the projector, transfers the presentation to the projection computer, and senses that the budget information should not be presented at that meeting—these also involve the use of smart spaces.

Aura uses cyber foraging to extend the capabilities of resource-constrained mobile devices by making use of compute servers or data staging servers that are connected to the Internet via a wired infrastructure. These servers are surrogates. Staging data on surrogates reduces the impact of end-to-end Internet delay. Although the use of caching can mask delay, cache misses may still happen, for example, because a mobile device does not have a cache large enough to store all relevant data. A cache miss may also occur when a file unexpectedly becomes relevant to the user; for example, the user receives a phone call from a client that requires the user to access a file that was previously considered unimportant.

Aura has a task layer, named Prism (Figure 8.2), that is a new abstraction to capture user intent. Prism is placed above individual applications and services. By introducing a new layer to explicitly represent user intent, the rest of the system has a basis on which to adapt or anticipate a user's needs (the example in scenario 1). In addition to supporting context observation that is appropriate for the environment, it also provides an environment management infrastructure that assists with resource monitoring and adaptation. These capabilities are encapsulated in the following components: task manager, context observer, and environment manager (Figure 8.3). The service supplier provides the support to carry out a user's task.

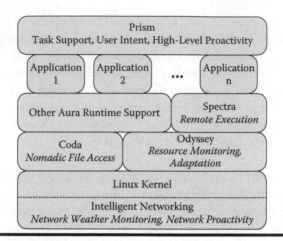

Figure 8.2   Aura architecture. (From Garlan, D., D. P. Siewiorek, A. Smailagic, and P. Steenkiste. 2002. Project Aura: Toward distraction-free pervasive computing. *IEEE Pervasive Computing* **1(2):22. Used with permission.)**

Figure 8.4 shows an application of Prism's architecture. Fred's home environment cooperates with his office environment. When Fred leaves home to go to work his home context observer alerts the home task manager. The task manager checkpoints the state of the running services platform-independently and causes the local environment manager to pause the services. This information and Fred's task state are stored in a distributed file space. When Fred enters his office environment, the office context observer detects his presence and informs the office task manager to reinstantiate the tasks by finding and configuring service suppliers in the new environment. The reconfiguration tries to maximize the use of local resources based on various resource utility functions specified by the task.

Prism's architectural framework provides a means to capture user intent to guide the search for suitable configurations in each environment. Tasks are rep-

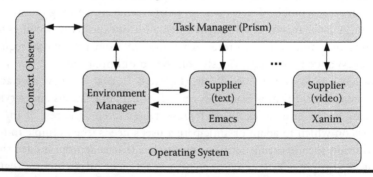

Figure 8.3   Prism architecture. (From Garlan, D., D. P. Siewiorek, A. Smailagic, and P. Steenkiste. 2002. Project Aura: Toward distraction-free pervasive computing. *IEEE Pervasive Computing* **1(2):22. Used with permission.)**

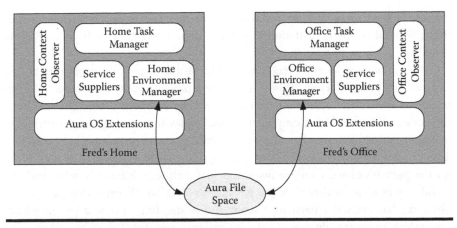

Figure 8.4   Aura task mobility. (From Garlan, D., D. P. Siewiorek, A. Smailagic, and P. Steenkiste. 2002. Project Aura: Toward distraction-free pervasive computing. *IEEE Pervasive Computing* **1(2):22. Used with permission.)**

resented as service coalitions. This helps Aura to determine when to support the essential services in a task and whether the services should be instantiated jointly. If a service cannot be provided, a user is given an early warning. The infrastructure then searches the heterogeneous environment for appropriate matches to provide the services. By providing the environment with self-monitoring capabilities, the infrastructure can detect when task requirements (such as minimum response time) are not met and can deploy alternative configurations to support the task. Aura is used to build the PHD discussed in chapter 1.

In context-aware applications such as HyperAudio and HIPS (Petrelli et al. 2001), described in Chapter 1, the challenge is deducing the next course of action for the system as the visitor moves to view the displays. If the visitor is detected standing in front of a certain display, the system presents information relevant to the display of interest. The components in the application that are responsible for deciding the visitor's context and what information to present are the input analyzer and the composer engine, which are rule-based modules. Rules are composed of a condition and an action. When the condition is satisfied, the rule is triggered and an action is performed. A condition is a set of tests on context variables that is continually updated. The context may be the visitor's position, the display he is looking at, or history information (e.g., information retrieved). Rules are applied to accomplish a certain task.

When a user approaches a new exhibit, the system reacts by sending a presentation, but this is not always the correct action. Consider the situation in which the visitor is listening to a presentation and moves toward a new exhibit. Should the system stop the current presentation or start a new one? The most appropriate action to take may depend on factors other than the visitor's movements, for example, the type of presentation currently running. If it concerns the object previously in focus it has to be interrupted or shortened because reference to the space is no longer valid. If the content of the presentation is general, then a full delivery is an acceptable behavior.

The use of context-awareness is helpful in minimizing error, for example, in a clinical application (Grasso 2004). To update a patient's data, the physician has to select the application, search for the correct patient and access relevant data about the patient. This would not only incur a small delay in accessing the patient's data but there is also a margin of error where a wrong patient's record might be selected. This process does not occur in a manual system where a physician simply picks up the patient's chart. Context-awareness may help eliminate this delay and the possibility of error by detecting the physician's location. When it detects that the physician has entered a particular room and is standing next to a patient's bed, the system automatically downloads the patient's relevant data to the physician's handheld. Location information can also be used as a passive security measure to deny access to data stored on the device when the device is not located within the vicinity of the hospital.

8.7.4 Mobile Work

When you work in an office, you work in a familiar environment. Your desktop is configured to suit your needs. You know where the photocopier and fax machine are located. If something breaks down, you know who to contact for support services. On the contrary, when you work away from your office, you are in an unfamiliar environment that may be very different from what you are used to. When you are at the airport, you cannot be certain how much bandwidth is available to you. If there are several users downloading large files, that might slow down your access to retrieve an important document that you need. You may have to make use of service discovery to find a printer near you.

When designing a mobile application or mobile device for use in any environment, it is imperative that it does not get in the way of work that is currently being carried out. The mobile application or device should be able to integrate into the existing environment without requiring its users to change the way they work significantly. Otherwise, the application or device would fail to meet its objective. The study discussed in section 8.7.1 illustrates the importance of understanding the users' working habits so that the application developed enhances the way they perform their tasks without requiring them to change the way they work significantly.

One of the objectives of mobility support is anywhere access. However, even though anywhere access is possible, it might not always be acceptable. For example, your colleague may be able to call you on your mobile phone, but you might not want to have the conversation in the presence of strangers who can overhear your discussion. Likewise, even though anytime access is possible, most people might not like to be contacted after what is an acceptable work hour.

In certain cases, knowing the user context is imperative before deciding to deliver particular information to the user. For example, while you are driving, you receive a call from a colleague who needs to clarify a few points regarding a document with you. For the communication to be effective, you need to see the document. Even though it is possible to deliver it to your device, it is not advisable as it may distract you from your driving.

Lack of awareness of a recipient's activity, time, and place may impact the rhythm of collaboration interaction between mobile workers and colleagues. For example, your colleague asked you to look at the attachment she e-mailed to you. You receive the e-mail but are unable to view the entire content due to the restrictions of your device. In this case, there is a mismatch in rhythm that results in problems in the ongoing collaboration activity.

One of the characteristic difficulties of mobile work is that there is less predictability and more restricted access to information and artifacts. Mobile workers do not know when they will have access to the Internet or what the quality of the connection will be when it is available. When traveling on business trips, mobile workers usually plan for all possibilities. In addition to bringing electronic documents, they also bring the hardcopy of the documents. Paper is flexible and is easier to access, whereas a laptop is unsuitable in certain circumstances. For example, when in a plane, reading paper documents is preferable and more convenient than accessing it from a laptop due to space constraints. A laptop and mobile phone can provide access to unanticipated documents. Mobile workers do not need to carry their offices with them: They just need to carry the means to access its functionality.

8.8   Summary

Developing mobile applications, especially in a smart environment, is more complex than developing traditional applications. Several additional issues need to be considered, and it requires a different way of thinking. Unlike the traditional environment where we can assume that users or applications will have a certain amount of resources available to them, similar guarantees may be invalid in a wireless environment as resources become available and disappear with time. The level or amount of resources available may also vary with time. Applications have to react correctly to these variations so that users can complete their tasks successfully. To accomplish this, support for adaptive behavior must be embedded in mobile application.

Heterogeneity is one of the important aspects that have to be kept in mind. The heterogeneity of devices makes it hard to design an application that is useful across different types of devices with various capabilities and functionalities. The type of data that can be delivered to the users must be tailored according to the formats that are supported by their devices. Data distillation is a technique used to deliver data in a format that is supported by a device while preserving most of the semantic content of data object by taking into consideration network, hardware and software constraints, and variations.

Alternatives to the conventional interface are needed because mobile users often perform multiple tasks. Interfaces that demand users' attention or that require a two-hand input are unsuitable as it may distract users from their tasks. Tactile interface is a means of providing an interface that is discreet and nonobtrusive. An alternative is a tilting interface where a device is equipped with a tilt or pressure sensor to provide a means of interaction that does not require two-handed input; for example, a user can scroll down a document by tilting a device. The scrolling stops when the device is returned to a neutral position.

Power conservation is important to prolong the duration between recharge. The strategies often involve powering down power hungry components such as the transceiver, display, hard disk, and CPU during idle periods. Applications should be designed to use power economically. In addition to power conservation, energy scavenging or harvesting offer attractive alternatives to powering up mobile devices. Currently, even though the amount of power that can be scavenged is on the order of milli- or microwatts, ongoing research in this area may offer better results in the future.

In the past decade, we have seen a transition in the way businesses are conducted from the physical stores to virtual stores on the Internet. The next transition is providing access to products and services via mobile devices. At the moment, mobile banking services are already available. One of the main concerns stated by users in a survey is security (Llazani and Othman 2004). Most respondents are skeptical about the security of conducting business transactions on their mobile devices. Constraints of processing power on mobile devices makes using strong encryption unviable.

Most mobile applications available today are targeted at teenagers and young professionals and are mostly infotainment applications, such as music download, friend locator, and news update. Mobile applications targeted for the enterprise environment that utilize smart space are not currently available in the market. Several issues need to be addressed satisfactorily to provide this type of application. We probably will have to wait a few more years to see the ubiquitous computing environment as envisioned by Mark Wiser becomes a reality.

References

Douglis, F., P. Krishnan, and B. Bershad. 1995. Adaptive disk spin-down policies for mobile computers. *Proceeding of the 2nd USENIX Symposium on Mobile and Location-Independent Computing*:121–137.

Eisenstein, J., J. Vanderdonckt, and A. Puerta. 2002. Adapting to mobile contexts with user-interface modeling. *Proceeding of the 3rd IEEE Workshop on Mobile Computing Systems and Applications*:83–92.

Fallman, D. 2002. Wear, point, and tilt: designing support for mobile service and maintenance in industrial settings. *Proceeding of the Conference on Designing Interactive Systems: Processes, Practices, Methods and Techniques (DIS 2002)*:293–302.

Flinn J., D. Narayanan, and M. Satyanarayanan. 2001. Self-tuned remote execution for pervasive computing. *Proceedings of the 8th Workshop on Hot Topics in Operating Systems*:61–66.

Flinn, J., and M. Satyanarayanan. 2000. Energy-aware adaptation for mobile applications. *ACM SIGOPS Operating Systems Review* 34(2):48–63.

Fox, A., S. D. Gribble, E. A. Brewer, and E. Amir. 1996. Adapting to network and client variability via on-demand dynamic distillation. *Proceedings of the 7th International Conference on Architectural Support for Programming Languages and Operating Systems* 31(5):160–170.

Garlan, D., D. P. Siewiorek, A. Smailagic, and P. Steenkiste. 2002. Project Aura: Toward distraction-free pervasive computing. *IEEE Pervasive Computing* 1(2):22.

Grasso, M. A. 2004. Clinical applications of handheld computers. *Proceedings of the 17th IEEE Symposium on Computer-Based Medical Systems* (CBMS 2004):141–146.

Llazani, A., and M. Othman. 2004. SMSBanking: An m-commerce application based on user requirement survey. *Proceedings of the 3rd Asian International Mobile Computing Conference 2004 (AMOC 2004)*:70–76.

Lee, Y. E., and I. Benbasat. 2003. Interface design for mobile commerce. *Communications of the ACM* 46(12):48.

Marsh, B., and B. Zenel. 1994. Power measurements of typical notebook computers. *Matsushita Information Technology Laboratory Technical Report MITL-TR-110–94*, May.

Noble, B. D., M. Satyanarayanan, D. Narayanan, J. E. Tilton, J. Flinn, and K. R. Walker. 1997. Agile application aware adaptation for mobility. *Proceedings of the 16th ACM Symposium on Operating System Principles:* 276–287.

Noble, B. D., and M. Satyanarayanan. 1999. Experience with adaptive mobile applications in Odyssey. *Mobile Networks and Application* 4(4):245.

Othman, M., and S. Hailes. 1998. Power conservation strategy for mobile computers using load sharing. *Mobile Computing and Communications Review* 2(1):44.

Paradiso, J. A., and T. Starner. 2005. Energy scavenging for mobile and wireless electronics. *IEEE Pervasive Computing* 4(1):18.

Petrelli, D., E. Not, M. Zancanaro, C. Strapparava, and O. Stock. 2001. Modeling and adapting to context. *Personal and Ubiquitous Computing* 5(1):20.

Poupyrev, I., S. Maruyama, and J. Rekimoto. 2002. Ambient touch: Designing tactile interfaces for handheld devices. *Proceedings of the 15th Annual ACM Symposium on User Interface Software & Technology 2002 (UIST '02)*:51–60.

Rudenko, A., P. Reiher, G. J. Popek, and G. H. Kuenning. 1998. Saving portable computer battery power through remote process execution. *Mobile Computing and Communications Review* 2(1):19.

Satyanarayanan, M. 2001. Pervasive computing: Vision and challenges. *IEEE Personal Communications*. 8(4):10.

Want, R., K. I. Farkas, and C. Narayanaswami. 2005. Energy harvesting and conservation. *IEEE Pervasive Computing* 4(1):17.

Weiser, M. 1991. The computer of the 21st century. *Scientific American*, September.

Bibliography

Bharghavan, V. 1995. Challenges and solutions to adaptive computing and seamless mobility over heterogeneous wireless networks. *Technical Report: Center for Reliable and High-Performance Computing*, Coordinated Sciences Laboratory, University of Illinois at Urbana-Champaign.

Imielinski, T., S. Viswanathan, and B. R. Badrinath. 1994. Energy efficient indexing on air. *Proceedings of the International Conference on Management of Data*, ACM SIGMOD:25–36.

Kakihara, M., and C. Serensen. 2002. Mobility: An extended perspective. *Proceedings of the 35th HICSS:*1756–1766.

Perry, M., K. O'Hara, A. Sellen, B. Brown, and R. Harper. 2001. Dealing with mobility: Understanding access anytime, anywhere. *ACM Transaction on Computer-Human Interaction* 8(4):323.

Satyanarayanan, M. 2004. The many faces of adaptation. *IEEE Pervasive Computing* 3(3):4.

Semrau, M., and A. Kraiss. 2001. Mobile commerce for financial services—Killer applications or dead end. *SIGGROUP Bulletin* 22(1):22.

Varshney, U., and R. Vetter. 2002. Mobile commerce: Framework, applications and networking support. *Mobile Networks and Applications* 7(3):185.

Online Resources

Mobile Information Access—Coda and Odyssey. http://www.cs.cmu.edu/~coda/ (Accessed February 16, 2007).

Project Aura. http://www-2.cs.cmu.edu/~aura/ (Accessed February 16, 2007).

Chapter 9

Location-Sensing and Location Systems

An important aspect of mobile computing is not only to allow users access to information anywhere, anytime but also to give information that is relevant to the user's context. In most cases, the user's location is an important factor in determining the user's context. For example, if I am in Seoul and I request for a list of restaurants, I would like to be presented with a list of restaurants near my vicinity, not a list of restaurants in Kuala Lumpur. For the system to deliver location-sensitive information, the system has to be able to detect my location and tailor the information delivered to me to meet my needs. This chapter discusses techniques to determine a user's location and how it is used in location-aware applications.

9.1 Location-Sensing Techniques

Let us start with definitions of the following terms:

■ Positioning systems: Provide the means to determine location and leave it up to users to compute their actual position. GPS is an example of a positioning system. There are two approaches to implement a positioning system: self-positioning and remote positioning. In a self-positioning system, a mobile device uses signals transmitted by gateways or antennas to calculate its own position. In a remote positioning system, the position of a mobile device or a tagged object is determined by measuring signals detected by a set of receivers. Signal measurements are used to determine the location of the object of interest.

■ Tracking systems: Monitor objects in their purview without involving the tracked objects in the computation.
■ Location sensing: A hybrid of the previous two systems, whether by design or configuration. The policy of manipulating location data is separate from the mechanism of pinpointing the object. Many systems are in this category.

There are three classifications of location-sensing techniques: triangulation, proximity, and scene analysis (Hightower and Borriello 2001).

9.1.1 Triangulation

Triangulation is divided into lateration and angulation. Lateration involves using distance measurements. The 2D position of an object is calculated by measuring its distance from three noncollinear points. Let us say you are lost somewhere and you have absolutely no idea where you are. You ask a passerby, "I'm lost. Where am I?" He replies, "You're 30 km from Bangi." You draw a circle on your map to mark your possible location. It is helpful but does not tell you exactly where you are because you could be anywhere within a 30-km radius from Bangi (Figure 9.1a). You ask another person the same question, and he says, "You're about 40 km from Kajang." That is slightly better. You can narrow down your location to the intersection area of the two circles (Figure 9.1b). A third person tells you that you are about 10 km from Serdang. You draw another circle on your map and the intersection point gives you your location—you are in Seri Kembangan (Figure 9.1c). This is how 2D lateration works.

Four non-coplanar points are required to calculate a 3D position (latitude, longitude, and altitude) of an object. A 2D position would be sufficient, for example, in an implementation involving an active badge that tracks the location of users or objects in an office environment. On the other hand, a 3D position would be more helpful when sending a team to rescue climbers trapped in an avalanche.

There are four approaches to measuring distance:

1. Direct: Involves using a physical action or movement to take a measurement; for example, a robot can take measurements using a tape measure. Obviously, this is not practical when dealing with roaming users and objects.
2. Time-of-flight: Involves measuring the distance between an object (stationary or moving) to a point P by measuring the time it takes to travel between the object and P at a known velocity. For example, because sound travels at a speed of 344 m/s, an ultrasound pulse transmitted by an object that arrives at P 14.5 ms later means that the object is 5 m away from P. If this technique is used, clock synchronization becomes an issue. If only one measurement is required, such as round trip time, the transmitter of the signal is also the receiver, and therefore, it needs to maintain its own clock with sufficient precision to calculate the distance. On the other hand, because a GPS receiver

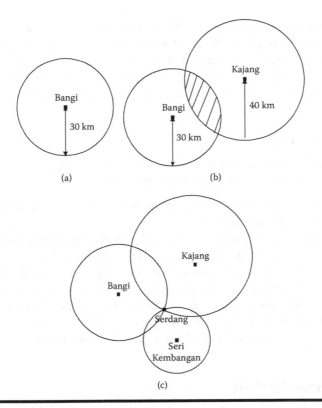

Figure 9.1   2D lateration.

is not synchronized with the satellites, it is unable to measure accurately the time a signal takes to reach it. Therefore, GPS satellites are precisely synchronized with each other and transmit their local time in the signal so that receivers can compute the difference in time-of-flight. A GPS receiver calculates a 3D position using four satellites. All of the 24 GPS satellites contain 4 cesium or rubidium atomic clocks that are locally averaged to maintain a time accuracy of 1 part in 10^{23} s.

3. Attenuation: The decrease in signal intensity as the distance from the signal source increases. It is possible to measure the distance between an object and P using a function that correlates attenuation, the distance for a type of emission, and the original strength of the emission. For example, a radio signal is attenuated by a factor proportional to $1/r^2$, where r is the distance from the source emitting the signal. Using attenuation is not as accurate as using time-of-flight due to obstructions in an indoor environment, reflection, refraction, and multipath propagation.

4. Angulation: Involves using angle or bearing measurements instead of distance. A 2D angulation involves two angle measurements and one length measurement as illustrated in Figure 9.2. On the other hand, a 3D angulation requires

one length measurement, one azimuth measurement, and two angle measurements. An example of the use of angulation is where phased antenna arrays comprising multiple antennas with known separation measure the arrival of a signal. The angle from the source of signal emission is calculated based on the differences in arrival times and the geometry of the receiving array.

9.1.2 Proximity

Proximity is a location-sensing technique that involves determining when an object is *near* a known location. There are three approaches:

1. Detection of physical contact using pressure sensors, touch sensors, and capacitive field detectors.
2. Monitoring wireless cellular APs involves detecting when a mobile device is within the range of one or more APs in a wireless cellular network.
3. Observing automatic ID systems involves using automatic ID systems such as credit card point-of-sale, telephone records, computer login history, and use of ATM card. For example, the location of a person driving on a highway can be inferred from the last time that person used a Touch-n-Go card because the scanner that reads the card has a static, known location.

9.1.3 Scene Analysis

Scene analysis uses features of a scene observed from a vantage point to deduce the location of the observer to objects in the scene. The observed scenes are simplified to obtain features that are easy to represent and compare. There are two types of scene analysis:

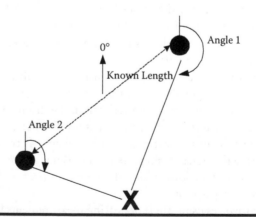

Figure 9.2 **2D angulation to locate an object X using angles relative to a 0-degree reference vector and the distance between two reference points.**

1. In static scene analysis, observed features are cross-referenced to a predefined dataset to map them to object locations.
2. In differential scene analysis, the differences between successive scenes are tracked to estimate location. The differences correspond to movements of the observer. If the features are identified as being at a specific location, the observer can determine its own position relative to them.

The scene may be visual images captured by a camera or other measurable physical phenomena, such as the electromagnetic characteristics that occur when an object is at a particular position and orientation.

9.2   A Taxonomy of Location Systems

Location systems are characterized by the following properties (Hightower and Borriello 2001b):

■ Physical and symbolic locations: A location system can provide two types of information: physical position and symbolic position. Physical position gives the absolute location of an object; for example, a GPS system gives the latitude, longitude, and altitude information to locate an object at its exact location. A symbolic position is relative, for example, in the lab on the second floor. A symbolic position is often based on proximity, such as a bar code scanner or an access card to a building. Depending on the requirements of an application, the location information may be physical or symbolic. For example, symbolic position is sufficient for an active badge. An application may also use both types of information. Symbolic position can be used to augment physical position; for example, a GPS coordinate (physical position) that determines a tourist's location is used to locate the nearest breakdown service center (symbolic position) for the tourist who is stranded on a highway.

■ Absolute or relative: An absolute location system uses a shared reference grid for all located objects; for example, a GPS system uses latitude, longitude, and altitude. On the other hand, a relative location system allows an object to have its own frame of reference, for example, a user's position to the nearest Thai restaurant.

■ Localized location computation: An object may compute its location locally. Local computation protects privacy, as no other object in the system may know the location of another object unless informed by the object itself. For example, GPS satellites have no knowledge of who makes use of the signal they transmit. On the contrary, an object may emit a signal that can be used by other objects to deduce its location. The signal may be broadcast periodically or in response to a request by another object. The infrastructure can then calculate the position of an object without involving the object in the computation. The advantage of this approach is the burden of performing the computation is placed on the infrastructure, hence, decreasing the computational and power demands on the object.

■ Accuracy and precision: Let us say a system is able to locate an object within a 10-m accuracy 95% of the time. The distance denotes accuracy (also termed grain size), and the percentage denotes precision (i.e., how often we can obtain that accuracy). Accuracy may be traded off for precision. Accuracy and precision may be improved using sensor fusion, where a number of location systems are integrated to form hierarchical and overlapping levels of resolution.

■ Scale: The scale of a location system refers to the coverage area per unit of infrastructure and the number of objects the system can locate per unit of infrastructure per time interval. For example, a location-sensing system may be able to locate objects within a room, a building, a campus, a city, or worldwide. A GPS satellite can serve an unlimited number of receivers, whereas an electronic tag reader is unable to read any tag if the distance is beyond its range. Time is an important factor in determining the scale of a location system as available bandwidth imposes a limit on the frequency and the number of location updates within a time interval. For example, if the number of communications in a RF-based system exceeds a certain threshold, the channel would become congested.

■ Recognition: Certain applications need to recognize objects to determine the next course of action. For example, a scanner reads the tag on luggage at an airport and sends it to the correct claim carousel; a system allows a user access to certain parts of the building after the fingerprint scanner identifies an authorized user. Among devices used by a system that incorporates recognition are cameras, scanners, and card readers. Generally, recognition techniques entail assigning a globally unique ID (GUID) to objects. The system looks up the GUID in a database to determine, for example, the object access rights to resources.

■ Cost: The cost of a location-sensing system includes time cost (time needed to install the system), space cost (the amount of installed infrastructure and hardware's size and form factor), and capital cost (price per unit or infrastructure element and salaries of support personnel).

■ Limitations: A system that functions well in one environment might not function as well in another environment. For example, IR systems are suitable for indoor use; a GPS system does not work indoors unless a GPS repeater is mounted on the building to rebroadcast the signals; certain tag readers can only read a tag correctly if there are no other tags nearby.

Table 9.1 gives a summary of location systems.

Location-Sensing and Location Systems 197

Table 9.1 Location-sensing technologies. (Hightower and Borriello 2001b.)

Technology	Technique	Physical	Symbolic	Absolute	Relative	Localized Location Computation	Recognition	Accuracy and Precision if Available	Scale	Cost	Limitations
Active Badges	Diffused IR cellular proximity		✓	◆			✓	Room size	1 base per room, badge per base per 10 s	Administration costs, cheap tags and bases	Sunlight and fluorescent light interfere with IR
Active Bats	Ultrasound time-of-flight lateration	✓		◆			✓	9 cm (95%)	1 base per 10 m2, 25 computations per room per second	Administration costs, cheap tags and sensors	Requires ceiling sensor grids
Automatic ID Systems	Physical contact proximity		✓	◇	◇		✓	Range of sensing phenomenon (RFID typically < 1 m)	Sensor per location	Installation, variable hardware costs	Must know sensor locations
Avalanche Transceivers	Radio signal strength proximity	✓			◆			Variable, 60- to 80-m range	1 transceiver per person	$200 per transceiver	Short radio range, unwanted signal attenuation
Cricket	Proximity, lateration		✓	◇	◇	✓		1.2 x 1.2 m regions (100%)	1 beacon per 1.5 m2	$10 beacons and receivers	No central management receiver computation
E911	Triangulation	✓		◆			✓	150 to 300 m (95%)	Density of cellular infrastructure	Upgrading phone hardware or cell infrastructure	Only where cell coverage exists
Easy Living	Vision, triangulation		✓	◆			✓	Variable	3 cameras per small room	Processing power, installation cameras	Ubiquitous public cameras

System	Technique						Accuracy	Scale	Cost	Limitations
GPS	Radio time-of-flight lateration	√		◆		√	1 to 5 m (95 to 99%)	24 satellites worldwide	Expensive infrastructure; $100 receivers	Not indoors
MotionStar	Scene analysis, lateration	√		◆			1 mm, 1 ms, 0.1° (nearly 100%)	Controller per scene, 108 sensors per scene	Controlled sensors, expensive hardware	Control unit tether, precise installation
MSR Radar	802.11 RF scene analysis and triangulation	√		◆			3 to 4.3 m (50%)	3 bases per floor	802.11 network installation, $100 wireless NICs	Wireless NICs required
PinPoint 3D-ID	RF lateration	√		◆			1 to 3 m	Several bases per building	Infrastructure installation, expensive hardware	Proprietary, 802.11 interference
Smart Floor	Physical contact proximity	√		◆			Spacing of pressure sensors (100%)	Complete sensor grid per floor	Installation of sensor grid, creation of footfall training dataset	Recognition may not scale to large population
SpotON	Ad hoc lateration	√			◊		Depends on cluster size	Cluster at least 2 tags	$30 per tag, no infrastructure	Attenuation less accurate than time-of-flight
VHF Omnidirectional Ranging	Angulation	√		◆		√	1-deg radial (100%)	Several transmitters per metropolitan area	Expensive infrastructure, inexpensive aircraft receivers	10 to 140 nautical miles, line of sight
Wireless Andrew	802.11 proximity		√	◆			802.11 cell size (100 m indoors, 1 km free space)	Many bases per campus	802.11 deployment, $100 wireless	Wireless NICs required, RF cell geometries

Note: ◊ means the system can be classified as either absolute or relative. ◆ means the system can only be classified as absolute or relative, depending on column marked. √ indicates property of the system.

9.3   GPS: An Example of a Positioning System

The GPS is a constellation of 24 satellites orbiting the earth. It was developed by the U.S. military as a military navigation system but has been made available to the public. The satellites are at a distance of 20,000 km above earth's surface. At any one time, four satellites cover a certain portion of the earth's surface. A GPS receiver determines its location based on signals it receives from the four satellites.

We gave an example of how 2D lateration works in section 9.1.1. A GPS receiver uses a 3D lateration to calculate its position in 3D—latitude, longitude, and altitude. Instead of three circles, imagine three spheres and their intersection point pinpoints your location. If you know that you are 1,000 km from Satellite 1, you could be anywhere on the surface of this imaginary sphere within a 10,000-km radius from the satellite. If you also know that you are 30,000 km from Satellite 2, you can narrow down your location to the intersection area of two spheres. If you know your location from a third satellite, the intersection point of the spheres on the earth's surface gives you your location. A GPS receiver usually uses signals from four satellites to improve accuracy and provide altitude information and can calculate its position with an accuracy of 10 m. How does a GPS receiver determine its distance from a satellite?

A GPS receiver picks up radio wave signals that travel at the speed of light from the satellites. The receiver determines its distance from the satellite based on how long a signal transmitted by the satellite takes to reach it. A satellite transmits a signal, termed a pseudorandom code, at predetermined intervals known by the receivers. When the satellite starts transmitting the code, the receiver starts to run the same code at exactly the same time. When the satellite's signal reaches the receiver, its signal pattern will lag slightly behind the code the receiver is running. The amount of time the signal lags behind (the delay) is equal to the signal's travel time. The receiver calculates its distance from the satellite by multiplying the delay by the speed of light (300,000 km/s). This calculation requires a receiver's clock to be synchronized with the satellite's clock.

DGPS is a technique to correct errors that may occur when a receiver calculates its location. A number of factors may cause the errors. First, an error occurs if the clocks are not perfectly synchronized. A receiver has to continuously reset its clock to be in sync with the satellite's clock. Second, for the distance information to be useful, a receiver has to know the position of the satellites. This information can be easily obtained because satellites travel in orbits. The receiver stores an almanac that tells it where a satellite should be at a given time. The orbits may vary slightly due to the pull of the sun and the moon, but the Department of Defense monitors the satellite's exact positions and transmits the adjustments to all receivers as part of the satellite's signal. Another source of error is the assumption that the signal travels at the speed of light. In reality, the signal is slowed down slightly as it enters the earth's atmosphere. This delay varies depending on where you are and cannot be factored in accurately when calculating the distance. Furthermore, radio signals are reflected off large objects, such as buildings, misleading the receiver into concluding that the satellite is farther away than its actual location. DGPS corrects these errors by estimating GPS inaccuracy at a stationary receiver station at a known location. Because this station knows its own location, it can easily calculate its receiver's inaccuracy. It then broadcasts a signal containing correction information to all DGPS-equipped receivers in its areas.

9.4   Active Badge: An Example of a Tracking System

In certain work places, such as a hospital, the ability to keep track of employees and equipment is critical for efficient operation. Conventional methods to locate an employee are using the phone to call at all locations, a pager, or a public-address system. These methods are time-consuming and inefficient. An alternative is to tag the employees or objects and track their movements automatically, which is what has been done at the Olivetti Research Lab (Want et al. 1992), where an active badge, with a size of 55 × 55 × 7 mm, is used as a tag. The badge emits a unique code (beacon) for 1/10 of a second every 15 s. Sensors placed throughout the buildings pick up the beacons.

What is the rationale behind the chosen signaling rate? Emitting and detecting signals consume power. Because power consumption is an important consideration, the signaling rate must be chosen carefully. The 15-s interval minimizes power consumption so that the battery can last for up to one year. Furthermore, the system must be able to detect several people in the same locality. Because the signal duration is 1/10 s, the probability that two signals will collide is about 1 in 150. Provided the number of people in the same locality is small, the probability that they will be detected is good.

Another power conservation strategy incorporated into the active badge is to turn off the badge when it is dark. As the level of lighting reduces, the period of the beacon signal is increased to an interval longer than 15 s. An ambient lighting environment alters the interval only slightly. If the badge is placed in the drawer during weekend or periods away from the office, the badge stops emitting the beacon and battery lifetime increases by a factor of four.

IR was used for signaling for a number of reasons. Because IR technology has been used commercially, it is inexpensive and readily available for new applications. IR transmitters and detectors are small, cheap, and can operate within a 6-m range. Sensors that are placed throughout the building detect the beacon and the information is used to identify the wearer's location. A master station polls the sensors for sightings of the badges, processes the data, and makes it available to clients.

A sensor network must be in place to provide thorough coverage. Sensors are placed high up on walls, ceiling tiles, entrances, and exits of corridors and other public areas. An active badge system can be implemented on a network of its own or integrated to an existing computer network. The latter approach decreases the cost of implementing the system significantly. The implementation at Olivetti Research Ltd. (ORL), England, consists of an independent network that supports 128 sensors. It allows workstations on an Ethernet network to support multiple badge networks. The data is relayed to the master station by conventional network protocols.

A badge network should be able to link all areas in a building with an arbitrary topology. Because there are many that are distributed in the building, possibly at remote places, power has to be fed through the network. Taking these factors into consideration, the badge-sensor network was designed as a four-wire network using conventional telephone twisted pairs. Two wires carry the network power supply, one carries the serial addressing information to allow the network controller to nominate a station, and the fourth carries data back to the network controller. To avoid requiring the network master to poll the sensors at high speed to avoid down-link, the first in, first out queue is capable of buffering 20 badge sightings. This allows the network master to multiplex between polling the network, manipulating data, and making the data available to clients.

The system maintains a table that consists of the name of the badge's owner, the nearest telephone extension, a description of the location, and the likelihood of finding that person at the location (given as a percentage). A likelihood of less than 100% indicates that the person is moving around. If he has not been sighted in the last five minutes, the field contains the last time and location he was sighted. This system is used by receptionists to locate employees.

The location server system consists of four layers:

1. Network controller: It is responsible for polling all sensors. Its polling strategy can be adapted to test sensors that have recently detected badges more frequently than those that have not. It detects errors in badge ID format and removes erroneous data.
2. Representation: Valid data extracted from the network is time-stamped and stored in a data structure that relates the badge ID to its location and the time it was sighted.
3. Data processing: An active badge network can collect potentially a large amount of data. The data should reflect only the changes in badge locations or provide compressed summaries a badge's recent history. The server was designed to compress the location information into a form that does not cause excessive traffic.
4. Display interface: The location information extracted from the other three layers is used as the input to a display function that shows textual information about a badge's changing location. Alternatively, a graphical interface may also be provided to the user.

Security and privacy are obvious concerns in using the system. Access to location data is restricted to ORL staff members only. If a person does not wish to be tracked, he can simply take off the badge and put it in his drawer or leave it on his desk. A user is allowed to specify who can locate them and has access to information about who has tried to locate them and how often.

9.5 Modeling Location-Tracking Application

Location-tracking applications provide a means to track elderly people, vehicles, merchandise, and other objects. For example, a company offering courier service may need to keep track of which vehicle is the nearest to a particular building to pick up a package. A taxicab company needs to know which of its drivers are within a five-minute distance to a certain address to pick up a customer. A car-for-hire company wants to keep track of the locations of its cars. An emergency service needs to know which of its ambulances is the closest to an accident scene. These examples involve human-controlled moving objects, also termed roving users (RUs).

Tracking the location of RUs can be achieved easily by using GPS. However, many applications require more than just simple tracking of users. Current technology offers a set of location-time pairs that shows the RU's location at a specific point in time. Due to technological constraints, the points in time may be minutes or hours apart, thus, it does not give accurate information of the location of a RU. This gives rise to the need to predict the location of users based on the last known location.

The techniques to predict location rely on estimates based on the object's direction, last reported location, and speed. This simple prediction technique is not always reliable, especially for real-life applications, because it assumes that speed and direction stay constant. This assumption may be valid only in certain situations.

A better approach is to manipulate the fact that humans are creatures of habit (Abdelsalam and Ebrahim 2004). Users have unique goals, personalities, and tasks. The same user might behave differently in different situations; hence, there is an element of uncertainty in predicting a user's location. One approach to predicting a user's location more accurately is by incorporating individual characteristics, habits, and preferences of a user. This involves building a user model that facilitates a more precise estimate of a user's location.

A user model aims to describe the cognitive processes that underlie the user's actions, the differences between the user's skill, the user's behavioral pattern or preferences, or the user's personal characteristics. For example, to reach a destination, different lorry drivers may prefer different routes (e.g., choice of route may be influenced by the time of the day), stop at different rest areas, or refuel at different petrol stations. Gathering this information about each lorry driver enables the location-tracking application to make more accurate predictions. The estimate made is no longer solely dependent on the driver's last known location and speed, but other unique characteristics that may influence the driver's decision.

Before building the user model, a set of variables has to be collected, depending on the type of application to be developed. The main types of variables are:

■ Temporal variables: Represent when an event occurs, for example, time of the day, day of the week.

  Spatial variables: Represent possible RU locations, for example, building, section of a town, highway.

  Environmental variables: Represent weather conditions, road conditions, and special events.

  Behavioral variables: Represent typical speed, resting patterns, and preferred rest area.

These variables are used to build a set of causality relationships between them. For example, there may be two routes to go from A to B. Joe prefers to take route 1, which is the shortest route but because route 1 is often very congested during rush hour, Joe takes route 2 if he has to make a delivery during this period. Therefore, there is a causality relationship between the route chosen and the time of the day.

To build and maintain a user model, data about the user's action must be collected continuously. Data may be collected periodically (e.g., every hour) or when a certain event happens (e.g., arrival at a rest area). The frequency of data collection depends on the application requirements. The more data is collected about the different variables, the more comprehensive and representative the model will become. The data collected is classified into two categories:

1. Environment-specific data describes different aspects of the environment, such as weather, traffic conditions, and special events taking place (e.g., certain roads are closed tomorrow due to the national day parade rehearsal).
2. User-specific data consists of personal or trip-related information.

The source of data depends on the data being collected; for example, user location is collected from a transmitter attached to the vehicle but weather condition is collected from a Web service. The collected data is used to update the user model automatically. An approach that can be used to build user models is a Bayesian network. In a Bayesian network, nodes are used to represent the variables—route, speed, time of the day, and weather conditions. Each node is associated with a probability that is calculated based on the user's history. Referring to Figure 9.3:

  Event: A list of common events that take place in the city.
  Time of day: May include rush hour, lunch hour, late night.
  Source: A list of sources for a trip (e.g., a building, street name, airport).
  Destination: A list of destinations for a trip.
  Weather conditions: Includes variables such as whether it is sunny, cloudy, raining.
  Route: A list of possible routes between a source and destination.

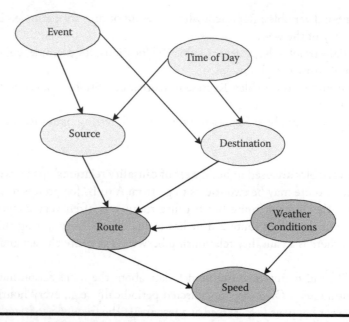

Figure 9.3 **A Bayesian network for a location-aware application. (From Abdel-salam, W., and Y. Ebrahim. 2004. Managing uncertainty: Modeling users in location-tracking applications.** *IEEE Pervasive Computing* **3(3):60. Used with permission.)**

■ Speed: All possible speeds.

Knowing the source, destination, and weather conditions enables us to predict the most probable speed. This information, together with the last known location, allows a prediction of a user's current or future location.

There are two operations that can be performed on a Bayesian network:

1. Maintain the network, which involves updating the probabilities associated with each node based on new observations
2. Perform inferences based on a given observation

The cost of the two operations must be examined to obtain an optimal performance. Updating the probabilities frequently will result in better prediction accuracy but may incur a high cost in terms of system performance. To lower the cost of maintaining the Bayesian network, we can lower the frequency of updates; for example, instead of making an update after each trip, the information is stored and updated once a week. To control the cost of maintenance and inference, the number of possible values at each node must be controlled. For example, instead of maintaining probabilities for each weather condition (e.g., sunny, cloudy, rain), we consider only good and bad weather. Keeping the user model small and simple results in easier updates and quicker response.

9.6 Location-Aware Application for Medical Workers

Hospital workers are constantly moving to perform their work. The information needs of medical workers depend on their locations and contextual conditions. Rodríguez et al. (2004) carried out a site study and identified the following information and services that are required by medical workers.

■ Patients' records, laboratory results, and forms need to be completed. A non-electronic method of handling these documents results in inconvenience when the documents are misplaced. Moreover, laboratory results may take up to eight hours to be delivered to the physician who requested them even though the processing of samples is mostly automated and takes only a few minutes to perform.

■ Patients and colleagues need to located. For example, a physician might need to locate a specialist to request an opinion.

■ Medical equipment, beds, and other devices are moved within the hospital as needed. There has to be a means of tracking and determining the availability of a required artifact quickly.

As the information and services required are often location-dependent, a context-aware hospital information system (HIS) that delivers information according to the user's location was developed after identifying the problems and requirements above. Rodríguez et al. (2004) presented the following scenario to illustrate the functionality required by the context-aware HIS:

> When Dr. Diaz is visiting a patient in bed 248, he realizes that he should request a lab test. He uses his PDA to request the test via the patient's electronic clinical record. When the chemist who is responsible for taking samples for the test visits the internal medicine area, his PDA informs him that a patient in bed 248 requires a lab test. When the chemist is standing at the patient's bed, his PDA lists the samples he must take in order to perform the analysis. After the analysis is done, he adds the results to the patient's record. When Dr. Diaz visits the patient in his next round, the result of the analysis is displayed on his PDA. Based on the result, he reevaluates the patient and decides to fill in a medical note to request the nurse in charge to increase the dose of the patient's medication.

The scenario shows how the system continuously estimates the location of the medical workers (in this case, Dr. Diaz and the chemist) to HIS. When they are near the patient's bed, relevant information to their role (a physician or a chemist) is displayed. The system is able to determine the identity of Dr Diaz and decides which forms to make available to him so that he can request a lab test. The system adapts to the context (in this case, location) of the user. The location-sensing technique used in this study is a back-propagation neural network that is trained to map RF signals from a WLAN to a 2D coordinate.

The location-aware HIS was developed as an agent-based system using a middleware named SALSA. A SALSA agent consists of three components:

1. A protocol to register the agent with an agent directory.
2. An instant messaging (IM) client. Users, user agents, and device agents interact via the IM client using eXtensible Markup Language (XML).
3. A subsystem that implements the agent's intelligence consists of a perception module, a reasoning module, and an action module. The perception module gathers knowledge through the IM client from sensors in the environment, other agents, or directly from users or services. The reasoning module determines an agent's actions, such as what to perceive next. The action module interacts with the environment by sending messages using a predefined protocol.

The interface of the application is based on the IM paradigm. Information relevant to the user's location and role is offered by default, but users can request other information by navigation or submitting a query. The handheld computers also provide forms that they can complete to communicate with their colleagues. The interface also notifies users of the presence and availability of other users and their locations. The information received in the handheld is retrieved from HIS. A HIS agent acts as a proxy of HIS and provides access to and monitors the state of the information in HIS. The architecture of HIS is shown in Figure 9.4.

Referring to the scenario given above, as Dr. Diaz visits his patients, the context-aware client in his PDA communicates with the location-estimation agent to update his position. When his location changes, the location-estimation agent sends an update of his location via the IM server to all users and agents who have

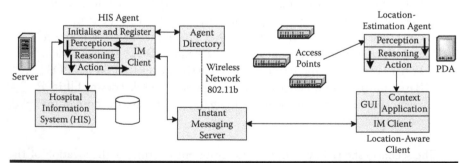

Figure 9.4   Architecture of the location-aware HIS. (From Rodríguez, M. D., J. Favela, E. A. Martínez, and A. Muñoz. 2004. Location-aware access to hospital information services. *IEEE Transactions on Information Technology in Biomedicine* 8(4):448. Used with permission.)

him registered in their roster. The HIS agent verifies if the contextual conditions match the new context (i.e., his role and location) to send him a message that he can use to retrieve the patient's record.

An important issue that is often debated when deploying location-aware applications is protecting the users' privacy. In the case of HIS, location estimation is performed at the handheld and location information is only shared with other users specified by the owner. This approach provides a measure of privacy protection. Because locating medical workers over the loudspeaker, by paging, and by sending messages to their mobile phone are common practices, the approach adopted by HIS does not pose a threat to their privacy. Furthermore, system configuration at the server can be done to prevent the sharing of location information when a user is in specified areas, such as the bathroom.

9.7   Summary

There are three methods that can be used to identify the 2D position (latitude and longitude) of an object: triangulation, proximity, and scene analysis. A 2D position is often sufficient for indoor applications, such as tracking of inventory in a warehouse or tracking of ME in a plant. On the other hand, a 3D position (latitude, longitude, and altitude) would be more helpful when dispatching a search and rescue team after receiving a distress call from a mountain climber. A popular means of identifying a 3D position is using a satellite service such as GPS.

A positioning system provides the means to determine location and leaves it to the user device to calculate its position, whereas a tracking system monitors objects without involving them in the computation. A location-sensing system is a hybrid of a positioning system and a tracking system.

A location-sensing system offers many advantages. One example is where a location system may prove critical in saving a person's life is the E911 Act, which specifies that a when person dials an emergency number, the network should make available the caller's number and location within a 125-m radius. This information is valuable for emergency services to locate a caller and dispatch help quickly.

Location systems often raise privacy concerns. For example, when the active badge was first introduced at ORL, it was met with great concern. To address security concerns, access to location data is restricted to ORL staff members only. Users are allowed to specify who can locate them. The management requested that all staff members wear the badge for a trial period of two weeks. During the trial period, incidence of missed telephone calls dropped substantially. Members of staff need not worry that they might miss an important call as calls are forwarded to where they are. Clients calling ORL got the impression of an efficient organization because the receptionist was able to locate a member of staff with certainty. Their receptionist can find a member of staff more quickly and informed visitors where they could find a person with greater confidence. Visitors were located easily when they were at ORL, which enhanced security and convenience. After seeing the benefits offered by the system, ORL staff continued wearing the badges after the trial period. Experience at ORL indicates that if users' privacy concerns are addressed appropriately and they are allowed to determine who can access their data, users are more willing to accept the implementation of a location system such as the active badge.

References

Abdelsalam, W., and Y. Ebrahim. 2004. Managing uncertainty: Modeling users in location-tracking applications. *IEEE Pervasive Computing* 3(3):60.

Hightower, J., and G. Borriello. 2001a. Location sensing techniques. *Technical Report UW-CSE-01–07–01,* Dept. of Computer Science and Engineering, University of Washington. Report. July 30.

Hightower, J., and G. Borriello. 2001b. Location systems for ubiquitous computing. *Computer* 34(8):57.

Rodríguez, M. D., J. Favela, E. A. Martínez, and A. Muñoz. 2004. Location-aware access to hospital information services. *IEEE Transactions on Information Technology in Biomedicine* 8(4):448.

Want, R., A. Hopper, V. Falcão, and J. Gibbons. 1992. The active badge location system. *ACM Transactions on Information Systems* 10(1):91.

Bibliography

Harter A., and A. Hopper. 1994. A distributed location system for the active office. *IEEE Network* 8(1):62.

Hightower, J., G. Borriello, and R. Want. 2000. SpotOn: An indoor 3D location sensing technology based on RF signal strength. *Technical Report UW-CSE-2000–02–02.* Dept. of Computer Science and Engineering, University of Washington. Report. February 18.

Hightower, J., and G. Borriello. 2001. A survey and taxonomy of location systems for ubiquitous systems. *Technical Report UW-CSE-01–08–03.* Dept. of Computer Science and Engineering, University of Washington. Report. August 24.

Scott, J., and M. Hazas. 2003. User-friendly surveying techniques for location-aware systems. *Proceedings of the 5th International Conference on Ubiquitous Computing, Lecture Notes in Computer Science* 2684:45.

Zeimpekis, V., G. M. Giaglis, and G. Lekakos. 2002. A taxonomy of indoor and outdoor positioning techniques for mobile location services. *ACM SIGecom Exchanges* 3(4):19.

Online Resources

The Global Positioning System. http://www.colorado.edu/geography/gcraft/notes/gps/gps_f.html (Accessed February 16, 2007).

Location Interoperability Forum (LIPF). http://www.cellular.co.za/technologies/location/lif.htm (Accessed February 16, 2007).

Location Pattern Matching Location Determination. http://www.911dispatch.com/911/lpm.html (Accessed February 16, 2007).

Chapter 10

Wireless Network Security

Co-authored by Mazliza Othman and Fazidah Othman

Wireless technologies have become increasingly prevalent in our lives. A PDA allows you to access calendar, e-mails, addresses, phone numbers, and the Internet. Some services offer GPS capabilities that can pinpoint the location of a device anywhere in the world. Wireless technologies promise to offer features and functions that are even more advanced in the next few years. The flexibility and cost savings have influenced organizations to deploy this technology without realizing that security risks in wireless networks are equal to the sum of operating a wired network plus the new risks introduced as a result of the portability of wireless devices. Issues such as transmission security and authentication are the most likely topics of discussion by security experts. At this point, encryption and authentication protocols are available to bring wireless network security to the same high level as that enjoyed by wired networks. Methods are available to prevent every known type of attack on wireless networks including WLAN detection, eavesdropping, MAC spoofing, rogue APs, theft of service, and potential legal issues. The only type of attack that can still threaten protected WLANs is one that also remains unresolved for wired networks: DoS, which is an attack that prevents legitimate users from accessing a network by flooding it with traffic. This chapter discusses security issues and threats that have not been covered in previous chapters.

10.1 Overview of Wireless Security Issues

Even though there are similarities between security issues in wired and wireless networks, the problems are aggravated by the fact that transmissions in wireless

211

networks are broadcasted over the air, making them more vulnerable to intercep-
tion and eavesdropping. Due to constraints in bandwidth, memory, and processing
power of mobile devices, more lightweight solutions are required when addressing
security concerns in wireless networks. The CPU and the transmitter are two components that consume significant amounts of power. Reducing energy consumption requires slower processing speed and processor cycle, thus, imposing a limit on
the complexity of encryption algorithms that can be executed on a mobile device.
Security protocols involve the sending and receiving of messages between parties
that are authenticating each other. Because transmission of data also consumes
much energy, the number of packets exchanged between the two parties must be
minimized or optimized.

There are four main security issues in wireless networks (Rahman and Imai
2002). The first issue is authentication. Mutual authentication is important and
beneficial for both service providers and users. It protects a service provider from
unauthorized intrusion and prevents a malicious entity from pretending to be a
BS. Mutual authentication allows a MS to authenticate a BS and then choose the
services of a particular BS in the presence of collocated networks.

There are a few issues related to authentication that have not been resolved
satisfactorily. Because most authentication protocols require that the authentication server at the user's home network be contacted for authentication purposes,
the server has to be available at all times. A part of the transparency requirement
of an authentication procedure is that the user should not notice a significantly
longer delay when the authentication takes place in a foreign network. This may
be difficult to meet as the completion time of the authentication also depends on
the quality of the link between the visited domain and the mobile user's home
authentication server (which might be unpredictable). The reliance on the home
authentication server may be eliminated by using certificates, but this approach is
not without its own set of problems.

First, no certification authority is globally accepted and trusted by all entities.
It is also impractical to assume that mobile users know in advance which certification authority is used by the domain they will be visiting. Second, a certificate says
nothing about the track record of the service provider in the foreign network nor
of the users. It does not offer complete protection to mobile users from malicious
entities in a foreign network. Conversely, the certificate says nothing about users'
behavior, such as their record in making payments for services rendered, which is
an important factor if the foreign network is going to charge roaming users for its
services. Even though it is possible to embed such information in a certificate, it is
hard to ensure that users cannot modify the information or present only certificates
that provide favorable credentials. Finally, regardless of whether the authentication
process is carried out with the users' home authentication servers or certification
authorities, the authentication process can be denied to users by cutting off the
communication channel between clients and the server or by flooding the network
so that no more bandwidth is available.

Even though mutual authentication protects the visited network and the user, it is still open to abuse. Users might not misuse services granted to them upon authentication to attack the visited network, but malicious users may connect to other organizations for the purposes of launching attacks and concealing their activities. To prevent this, a service provider must have a means to detect suspicious activities and halt them. Failure to do so might be costly to the service provider's reputation if it is identified to be the source of an attack.

The second issue involves privacy and anonymity. Anonymity involves keeping private a user's real identity, activities, current location, and movement patterns. Guaranteeing anonymity is harder in a wireless network because it is more susceptible to eavesdropping and tapping. Protecting a user's location and movement pattern is important because it may provide an insight to a user's activity and who the user interacts with at a given time. The following precautions were proposed to ensure anonymity:

■ Prevent any association of users with messages sent or received.
■ Prevent any association of users with communication sessions that they participate in.
■ Preserve the privacy of location and movement information of users.
■ Prevent the disclosure of the relationships between users and their home domains.
■ Prevent any association of users with foreign domains they visited.
■ Avoid exposing users' activities by hiding the relationships between them and visited domains.
■ Use aliases or nicknames (short- or long-lived) to refer to users without revealing their real identities. A short-lived alias changes each time an intercell or interdomain handoff occurs, whereas a long-lived alias is valid for a predefined period.

A DoS involves overwhelming a server with requests to deny legitimate users access to data and services offered by the server. It is harder to control visiting hosts from overloading the network with excessive transmission, resulting in a sudden decrease in network performance, which may lead to a DoS to other mobile hosts. A DoS attack is classified as an unselective DoS or a selective DoS. Unselective DoS disables the entire or a large part of a network. On the other hand, a selective DoS is less evident and its victims are usually well-defined. Ensuring anonymity would make selective DoS ineffective.

The third issue is device vulnerability. The portability of mobile devices makes it easier for them to be misplaced, lost, or stolen. Even if no confidential data is stored on the device, the loss of contact information, for example, can be costly in terms of inconvenience and the time a user takes to recover the data. On the other hand, data that the user considers unimportant might be valuable to the thief. A mobile device such as an active badge can be used as a control device that gives users access to services and information. A thief who can disable the device's security features will have access to information and services that are granted via the device. Passwords offer poor protection. One way to overcome this problem is to encrypt data stored on devices, such as smart cards, at the expense of battery power and processing capability.

The fourth issue is domain crossing. A security domain consists of a set of network entities under one administrative authority employing one security policy. When mobile users enter a new domain, they need to authenticate each other to ascertain the trustworthiness of one another. The level of trust then determines what services and information the new domain wants to offer to the mobile users and what activities the mobile users consider safe to carry out while in the domain. Another reason why the authentication process is important to a domain is so that it can uphold its image as a safe domain. If a domain that sells its services gains a reputation of being unsafe, mobile users may choose to steer away from it, possibly resulting in lost revenue for the domain.

10.2 Security of Data Transmission

The objectives of security mechanisms and protocols are to ensure the privacy and integrity of data, the authenticity of communicating parties, provide nonrepudiation, prevent DoS attacks, filter viruses and malicious code, and support communication anonymity. Data transmission security is divided into four domains (Ravi et al. 2002):

1. Appliance domain security ensures that only authorized entities can access an appliance and modify the data stored on it.
2. Network access domain security ensures that only authorized devices can connect to a wireless network and service and guarantee data privacy and integrity over the wireless link
3. Network domain security deals with the security of the infrastructure that supports a wireless network and may span wired or public networks and may be owned by multiple carriers.
4. Application domain security ensures that only safe and trusted applications can execute on the appliance, and that transactions between client and server applications across the Internet are secure.

Many existing security protocols, such as security protocols defined for WLAN, Bluetooth, and GSM, address only the network access domain security (i.e., securing the link between a wireless client and the AP, BS, or gateway). The protocols used in bearer technologies are insufficient for data requiring high level security and do not address the problem of maintaining end-to-end security across the wired infrastructure network.

Addressing security concerns for wireless clients is more challenging due to their relatively low processing capability and dependence on battery power. Additionally, mobile devices are susceptible to theft, and this is another factor to consider to secure data stored on mobile devices. The following challenges must be met when securing wireless devices and communications (Ravi et al. 2002):

- Security processing gap: Security processing requirements put a very high demand on the limited resources of wireless clients, significantly impairing user experience due to poor connection latencies, lower data rates, and shorter battery life. As an example, PalmIIIx requires 3.4 min to perform a 512-bit RSA key generation and takes 7 s to perform digital signature generation.
- Battery gap: The growth of battery capacity at 5–8% a year is much slower than the security processing power requirement. The increase in energy requirements of wireless devices is outpacing the growth in battery capacities. The addition of security will further widen the battery gap, for example, a Sensoria WINS WSN node requires 21.5 mJ to transmit a 1024-bit message, but encrypting the same message using RSA requires 42 mJ.
- Flexibility: A security protocol standard specifies a wide range of cryptographic algorithms for a client and server to execute to facilitate interoperability. A wireless device often has to execute multiple distinct security protocol standards to support security processing at different layers of the network protocol stack and for interworking among different networks. Efficiency of security processing has to be considered with, and traded off against, the need for flexibility. Furthermore, ease of upgrade is desirable to ensure smooth upgrades in the future as the security standards evolve.
- Tamper-proof implementation: This is important to minimize the security risk if a mobile device physically falls into the hands of a malicious entity. A simple password or PIN protection is not sufficient to prevent unauthorized access. On the other hand, incorporating biometric techniques introduces a set of new challenges, such as how do we integrate a fingerprint scanner to an already constrained mobile device.

Ravi et al. (2002) show that there is a wireless security processing gap due to a mismatch between wireless security processing requirements and the capabilities of processors in mobile devices. They propose the following approaches to bridge the gap:

- Low complexity security protocols and cryptographic algorithm: To maintain interoperability between wireless devices and their wired peers, many existing wired protocols need to be supported on the wireless device. The security processing workload on a mobile client can be reduced by carefully selecting and implementing only a subset of a protocol's features. A security protocol can be made to adapt its encryption policy based on the type of data being encrypted. For example, a video encryption algorithm protects only the more important parts of a video stream instead of the entire stream to reduce the amount of data encrypted. The use of lightweight algorithms would also contribute to reducing the demand on processor capability. The ECC is an example of a public-key cryptosystem that provides a high level of security and is less demanding on computing and memory resources.

■ Embedded processors with enhanced security processing capabilities: These involve adding cryptographic accelerators to the basic processing core by accelerating bit-level arithmetic operations for permutations performed using Data Encryption Standard (DES) or 3DES. Another approach is using instruction set extensions for substitutions, rotations, and modular arithmetic operations that are used in private-key algorithms.

■ MObile SEcurity processing System (MOSES): A programmable security processor platform developed at NEC to secure data and multimedia communications in next-generation wireless devices. Its objective is to address the wireless security processing gap and to support a wide range of current and future security protocol standards. The MOSES system architecture consists of layered, optimized software libraries that implement the cryptographic algorithm and a state-of-the-art configurable and extensible processor that is customized for efficient security processing.

10.3 Next-Generation Hackers

Virus, worm, and Trojan attacks are classified under malware threats. Mobile phones have evolved to become sophisticated full-scale Internet-enabled computers. In the next few years, attacks on mobile phones and devices will become similar to the ones on PCs. Dagon et al. (2004) introduce a taxonomy of mobile malware based on the type of security that attackers aim to compromise (Table 10.1). There are three types of attacks: confidentiality, integrity, and availability attacks.

Information theft can be classified into two categories (Dagon et al. 2004): transient information and static information. Examples of transient information are the phone's location, its power usage, and other data the device does not record. A user's movement can be tracked using the E911 service. Static information is information that mobile devices store and transmit over the network, such as phone numbers and programs stored on the device. Bluesnarfing is an attack that targets static information. It gives hackers access to data stored on a Bluetooth-enabled phone using Bluetooth wireless technology without alerting the phone's owner of the connection made to the device. The hacker would be able to access the owner's phonebook, images, calendar, and IMEI. This attack can be prevented by setting the device to nondiscoverable mode.

Table 10.1 Mobile malware taxonomy.	
Compromised Security Goal	*Examples of Attacks*
Confidentiality	Theft of data, bluebugging, bluesnarfing
Integrity	Phone hijacking
Availability	Protocol-based DoS attacks, battery draining
Source: Dagon, D., T. Martin, and T. Starner. 2004. Mobile phones as computing devices: The viruses are coming! *IEEE Pervasive Computing* 3(4):11. Used with permission.	

Bluebugging is an attack that gives the hacker access to mobile phone commands using Bluetooth wireless technology without notifying or alerting the phone's owner. This vulnerability allows the hacker to initiate phone calls, send and read SMS, read and write phonebook contacts, eavesdrop on phone conversations, and connect to the Internet. Most people believe that because Bluetooth is a short-range communication standard, threats to mobile phones are minimal as the attackers have to be physically near their victims, but it has been shown that it is possible to establish a Bluetooth connection with a standard mobile phone almost 2 km away using a 19-dbi (decibels relative to isotropic radiator) panel antenna (G4 Media Inc. Web site). Bluebugging gives the attacker access to the victim's phone and turns it into a listening device. The attacker can call the victim's device without ringing it and the phone automatically answers. The attacker turns the victim's phone into a bugging device and eavesdrops on the owner's conversation. Phone jacking is a malware that allows an attacker to use the victim's phone resources to place long distance call and send expensive MMS messages.

There are two types of DoS attacks. The first type tries to flood the device, and the other one tries to drain the battery. Open-source Bluetooth stacks are vulnerable to simple protocol attacks such as repeated sending of packets and sending incorrect packet formats. The battery is drained by preventing it from going into PSM. This means that a mobile phone's battery would only last as long as the talk time even if the user is not using the phone. If the battery-draining malware could spread quickly from one mobile phone to another, the attacker could disrupt mobile phone service on a large scale.

Security researchers have identified a number of ways hackers can infect mobile phones with malware (Leavitt 2005). Smart phones are most susceptible to malware threats because they provide users with Internet connectivity that allows users to download applications and files that may contain malicious code. Considering that an estimated 130 million smart phones will be sold by 2008, the scale of damage that can be done by the next-generation hackers may potentially be devastating. Amateur virus writers may cause minor nuisances such as deleting a victim's personal data, but professional hackers might do more serious damage, for example, create an attack that degrades or overloads mobile networks. Furthermore, as m-commerce gains acceptance and malware becomes more sophisticated, an attacker would be able to steal financial data, such as a credit card number stored on the phone. Viruses can allow an attacker access to passwords and corporate data stored on the phone. The victim's phone can also be manipulated to make calls and send messages, running up the victim's phone bill. This attack is termed theft of service. The first mobile virus was found in 2000 and about 31 viruses have been detected by 2005. Table 10.2 lists a few of the identified mobile malicious codes.

SMS with its limit on 168 characters is not useful as a means to spread viruses but can be used to generate a huge amount of SMS traffic that could overwhelm the network or incur high bills for the victims. Future viruses are likely to be spread by MMS. A MMS that can carry up to 150 kbps of data is large enough to be a means to spread viruses. Even though many mobile phones now support e-mail applications, it is unlikely to become a means to transmit mobile viruses as users rarely use their phones exclusively to read e-mails. Virus writers prefer to send malicious code using approaches primarily used by mobile phones. Therefore, as the popularity of mobile IM grows, attacks similar to the one seen on desktops, such as hijacking an IM name list, are likely to increase. Another means of transmitting mobile code is placing infected mobile games on the Internet for downloading.

10.4 Summary

The security risks in wireless networks are equal to the sum of risks when operating a wired network plus the risks imposed by the vulnerability of wireless networks and the portability of mobile devices. The security threats are exacerbated due to the vulnerability of transmitting data over the air. As technology progresses, we will see more desktop-like applications and services offered to mobile users. Some of these services and applications may require the transmission of confidential or sensitive data, such as a credit card number or a patient's medical data. Encryption of such data proves to be a delicate balancing act between meeting the demand of a processor-intensive encryption algorithm and conserving power on a resource-constrained mobile device.

The provision of a seamless service to users as they roam between domains is a desirable aspect of mobile service but brings its own set of security challenges. Mutual authentication is a must to protect both users and service providers from malicious entities. Upon authentication, a user is assigned a trust level that determines the types of services and resources that the user can access.

Location tracking is necessary to provide services and context-relevant information to mobile users. Because the location-tracking information can easily be misused to track a user's habits and activities, it is imperative that procedures and policies are in place to protect users' privacy and anonymity.

Table 10.2 Malicious Code.

Name	Type	Spread by	Attack
Cabir	Worm	A shared infected application, Bluetooth link.	The worm arrives as a .SIS (Symbian installation system) application installation file. The victim's device asks if the user wants to receive a message via Bluetooth. If the user chooses to receive the file, it installs and sends itself to another Bluetooth-enabled device. It forces the device to constantly scan for other Bluetooth-enabled devices, reduces battery lifetime, and degrades Bluetooth performance. This attack can be prevented by turning off the Bluetooth setting on the phone.
Skulls	Trojan horse	Hackers upload Skull to shareware sites where it is downloaded by users. It appears as a useful application that lets users preview, select, and remove design themes on their hand phones.	It disables Symbian applications by replacing the original binaries, such as functions for file management, messaging, and Web browsing, with nonfunctional binaries. As a result, the phone can only be used to make and receive calls. It replaces application icons with a skull and crossbones.
Mquito	Worm	Downloaded file.	It is a version of the Mosquito game whose copy protection has been cracked. When the game is installed, it sends unauthorized SMS text messages to high-cost toll phone numbers in Germany, Holland, Switzerland, and the United Kingdom.
Windows CE Virus	Virus	It spreads by file exchange or transfer. The virus sends a message asking for permission to download.	If the victim allows the download, the virus infects all executable files bigger than 4 KB by appending itself to the file. When the victim tries to execute the file, the virus runs, but not the application. This virus was written as a proof-of-concept for Microsoft mobile operating system.
Metal Gear	Trojan horse	The Trojan horse is camouflaged as a Metal Gear Solid video game. It spreads when a victim opens and installs a fake Metal Gear game.	It disables antivirus programs and installs the Cabir.G worm that tries to spread another Trojan, SEXXXY, to nearby phones via Bluetooth links. It disables all tools on the phone that are needed to undo the damage done.
Gavno	Trojan horse	It contains an application file that has been rendered invalid by hackers.	When the Symbian operating system tries to use the file, it causes a series of cascading errors, making the operating system unstable. The infected phone can only receive calls. Gavno makes the phone reboot, producing similar errors.

Source: Leavitt, N. 2005. Mobile phones: The next frontier for hackers? *Computer* 38(4):20. Used with permission.

220   *Principles of Mobile Computing and Communications*

Users can protect themselves from malware threats by following the simple guidelines that protect them when on the wired networks, such as not downloading files distributed at Web sites and not opening file attachments. Attacks that manipulate Bluetooth security vulnerabilities can easily be avoided by setting the device to nondiscoverable mode.

References

Dagon, D., T. Martin, and T. Starner. 2004. Mobile phones as computing devices: The viruses are coming! *IEEE Pervasive Computing* 3(4):11.

Leavitt, N. 2005. Mobile phones: The next frontier for hackers? *Computer* 38(4):20.

Rahman, M. G., and H. Imai. 2002. Security in wireless communications. *Wireless Personal Communications* 22(2):213.

Ravi, S., A. Raghunathan, and N. Potlapally. 2002. Securing wireless data: system architecture challenges. *Proceedings of the 15th International Symposium on System Synthesis*:195–200

Bibliography

Ashley, P., H. Hinton, and M. Vandenwauver. 2005. Wired versus wireless security: The internet, WAP and iMode for e-commerce. *Proceedings of the 17th Annual Computer Security Applications Conference 2001 (ACSAC 2001)*:286–306.

Hole, K. J., E. Dyrnes, and P. Thorsheim. 2005. Securing Wi-Fi networks. *Computer* 38(7):28.

Johnston, D., and J. Walker. 2004. Overview of IEEE 802.16 security. *IEEE Security & Privacy* 2(3):40.

Manley, M. E., C. A. McEntee, A. M. Molet, and J. S. Park. 2005. Wireless security policy development for sensitive organizations. *Proceedings of the 2005 IEEE Workshop on Information Assurance*:150–157.

Welch, D., and S. Lathrop. 2003. Wireless security threat taxonomy. *Proceedings of the 2003 IEEE Workshop on Information Assurance*:76–83.

Online Resources

802.11 Security Vulnerabilities. http://www.cs.umd.edu/~waa/wireless.html (Accessed February 16, 2007).

G4 Media, Inc. http://www.g4tv.com/screensavers/features/48021/Bluetooth_Attack.html (Accessed December 9, 2005).

The Unofficial 802.11 Security Web Page. http://www.drizzle.com/~aboba/IEEE/ (Accessed February 16, 2007).

Acronyms

1G	first generation
2D	two-dimensional
2G	second generation
2.5G	second-and-a-half generation
3D	three-dimensional
3G	third generation
3GPP	Third Generation Partnership Project
AAA	accounting, authentication, and authorization
ACH	access feedback control channel
ACK	acknowledgment
ACL	asynchronous connectionless link
ACM	address complete message
AES	advanced encryption standard
AI	acquisition indicator
AICH	acquisition indicator channel
AIFS	arbitration interframe space
AIO	abstract interaction object
AM	active mode
AMS	advanced mobile phone system
AN-SN	access network–serving network
ANSN	advertised neighbor sequence number
AODV	ad hoc on-demand distance vector
AP	access point; access preamble
AP-AICH	CPCH access preamble acquision indicator channel
API	access preamble acquisition indicator

APS	application support
AR	access router
ARP	Address Resolution Protocol
ARQ	automatic repeat request
ASCENT	Adaptive Self-Configuring sEnsor Networks Topologies
ATIM	announcement traffic indication message
AuC	authentication center
B2B	business-to-business
B2C	business-to-customer
BCCH	broadcast control channel
BCH	broadcast channel
BPSK	binary phase shift keying
BR	border router
BRAN	broadband radio access networks
BS	base station
BSC	base station controller
BSS	base station subsystem; basic service set
BSSAP	base station system application part
BTS	base transceiver station
CAP	contention access period
CD/CA-ICH	CPCH collision detection/channel assignment indicator channel
CDI	collision detection indicator
CDMA	code division multiple access
CFP	contention-free period
CIO	concrete interaction object
CIP	cellular IP
CIR	carrier-to-interference ratio
CML	client modification log
COA	care of address
COTS	commercial off-the-shelf
CP	contention period
CPCH	common packet channel

CPICH	common pilot channel
CPU	central processing unit
CRC	cyclic redundancy check
CSICH	CPCH status indicator channel
CSMA	carrier sense multiple access
CTS	clear to send
CW	contention window
DCF	distributed coordination function
DCH	dedicated channel
DECT	digital enhanced cordless telecommunica-tion
DES	Data Encryption Standard
DFS	dynamic frequency selection
DGPS	differential global positioning system
DHCP	Dynamic Host Configuration Protocol
DIFS	distributed coordination function interframe space
DiL	direct link
DL	data loss; downlink
DLC	data link control
DLL	data link layer
DOCSIS	Data Over Cable Service Interface Specification
DoS	denial of service
DPCCH	dedicated physical control channel
DPCH	dedicated physical channel
DPDCH	dedicated physical data channel
DSCH	downlink shared channel
DSR	dynamic source routing
DSSS	digital sequence spread spectrum
DTN	disruption-tolerant networking
EAP	Extensible Authentication Protocol
ECC	elliptic curve cryptography
ECG	electrocardiogram
EDCF	enhanced distributed coordination function
EDGE	Enhanced Data Rates for GSM Evolution

EIR	equipment identity register
ETSI	European Telecommunications Standard Institute
EWG	enterprise wireless gateway
FACH	forward access channel
FCC	forward control channel
FCH	frame control channel
FEC	forward error correction
FFD	full function device
FHSS	frequency hopping spread spectrum
GCR	group call register
GFSK	Gaussian-shaped frequency shift keying
GGSN	General Packet Radio Services gateway support node
GMSC	gateway mobile-services switching center
GPRS	General Packet Radio Service
GPS	global positioning system
GRE	generic router encapsulation
GSM	Global System for Mobile Communications
GTS	guaranteed time slot
GUI	graphical user interface
GUID	globally unique ID
HC	hybrid coordinator
HCF	hybrid coordination function
HCI	host controller interface
HIS	hospital information system
HLR	home location register
HMIPv6	Hierarchical Mobile Internet Protocol version 6
HRFWG	HomeRF Working Group
HSCSD	High-Speed Circuit-Switched Data
HS-DSCH	high-speed downlink shared channel
HSEQ	HELLO sequence number
HS-PDSCH	high-speed physical downlink shared channel
HS-SCCH	shared control channel
IAM	initial address message

IC	integrity code
ICMP	Internet Control Message Protocol
IEEE	Institute of Electrical and Electronic Engineers
IETF	Internet Engineering Task Force
IFS	interframe space
IM	instant messaging
IMEI	international mobile station equipment identity
IMSI	international mobile subscriber identity
I/O	input/output
IP	Internet Protocol
IR	infrared
ISDN	International Services Digital Network
ISM	industrial, scientific, and medical
ITU	International Telecommunication Union
IV	initialization vector
L2CAP	Logical Link Control Adaptation Layer Protocol
LA	location area
LAI	location area identification
LAN	local area network
LCH	long transport channel
LE	local exchange
LLC	logical link control
LRU	least recently used
LR-WPAN	low-rate wireless personal area network
LT	loss threshold
LW	logical window
MAC	medium access control
MANET	mobile ad hoc network
MANNA	Map ANNotation Assistant
MAP	mobile application part
MD5	message digest 5
ME	mobile equipment
MISDN	mobile station international ISDN number
MMS	multimedia message service

MOSES	MObile SEcurity processing System
MPDU	MAC protocol data unit
MPEG	Motion Picture Experts Group
MPR	multipoint relay
MS	mobile station
MSC	mobile switching center
MSDU	MAC service data unit
MSISDN	Mobile Station International Services Digital Network
MT	mobile terminal
MT-AN	mobile termination–access network
MT-SN	mobile termination–serving network
NAV	network allocation vector
NCM	network connection monitor
NIC	network interface card
NT	neighbor threshold
OFDM	orthogonal frequency division multiplexing
OLSR	Optimized Link State Routing
O-QPSK	offset quadrature phase shift keying
ORL	Olivetti Research Ltd.
OSPF	Open Shortest Path First
PAN	personal area network
P-CCPH	primary common control physical channel
PC	personal computer
PCF	point coordination function
PCH	paging channel
PCPCH	physical common packet channel
PCS	personal communication service
PDA	personal digital assistant
PDP	Packet Data Protocol
PDSCH	physical downlink shared channel
PDU	protocol data unit
PHD	Portable Help Desk
PHY	physical

PICH	paging indicator channel
PIFS	point coordination function interframe space
PIN	personal identification number
PLMN	public land mobile network
POS	personal operating space
PPDU	physical protocol data unit
PPP	Point-to-Point Protocol
PRACH	physical random access channel
PRNG	pseudo-random number generator
PS	power save
PSDU	physical service data unit
PSM	power-saving mode
PSTN	public switched telephone network
PTIP	Periodic Terminal Initiated Polling
PU	presentation unit
QAM	quadrature amplitude modulation
QoS	quality of service
QPSK	quadrature phase shift keying
RAB	radio access bearer
RACH	random access channel
RADIUS	remote authentication dial-in user service
RAN	radio access network
RCC	reverse control channel
RC4	Rivest Cipher 4
RDP	Resource Discovery Protocol
RF	radio frequency
RFCOMM	radio frequency communications
RFD	reduced function device
RFID	radio frequency identification tag
RID	router identification
RNC	radio network controller
RNS	radio network subsystem
RR	resource request
RRC	radio resource control

RREP	route reply
RREQ	route request
RRER	route error
RSSI	received signal strength indication
RT	reported tree
RTC	real-time clock
RTS	request to send
RTT	radio transmission technology
RU	roving user
SAR	segmentation and reassembly
SC	service center
S-CCPH	secondary common control physical channel
SCH	short transport channel; synchronization channel
SCO	synchronous connection-oriented link
SDP	Service Discovery Protocol
SDU	service data unit
SGSN	serving gateway support node
SIFS	short interframe space
SIS	Symbian installation system
SMS	Short Message Service
SN	sequence number
SN-HN	serving network–home network
SOHO	small office / home office
SP	signalling point
SPI	security parameter index
SRNS	serving radio network subsystem
SS7	Signalling System No. 7
SSID	service set identifier
SUI	speech user interface
SWAP	Shared Wireless Access Protocol
TBRPF	Topology Broadcast based on Reverse Path Forwarding
TBTT	target beacon transition time
TC	topology control; traffic category
TCP	Transmission Control Protocol

TCS	telephony control specification
TDD	time division duplex
TDMA	time division multiple access
TE-MT	terminal equipment–mobile termination
TG	topology graph
TIM	traffic indication map
TKIP	Temporal Key Integrity Protocol
TND	Topology Broadcast based on Reverse Path Forwarding neighbor discovery
TT	topology table
TTL	time-to-live
TXOP	transmission opportunity
UDP	User Datagram Protocol
UE	user equipment
UI	user interface
UL	uplink
UMTS	Universal Mobile Telecommunications Service
USIM	user service identity module
UTRAN	UMTS Terrestrial Radio Access Network
VLR	visitor location register
VOIP	Voice over IP
WCAC	wireless connection admission control
WCC	wireless congestion control
WCDMA	wideband code division multiple access
WEP	Wired Equivalent Privacy
Wi-Fi	wireless fidelity
WiseMAC	Wireless Sensor medium access
WLAN	wireless local area network
WPAN	wireless personal area network
WSN	wireless sensor network

Index

NOTE: Page numbers in **bold** are figures and tables.

Number

128-bit Data encryption, 52
1G (*See* first generation)
1xRTT (*See* radio transmission technology standard)
2.4 GHz, 5, **88**, **91**
2.5G systems, 7
2D angulation, 193, **194**
2D position, 192
2G (*See* second generation)
3D (Three-dimensional), 37
3D angulation, 193
3DES, 216
3D lateration, 199
3D position, 192–194, 199, 207
3G (*See* third generation)
3G networks, 38–39
3G systems, 7
3GPP TS 23.101 V5.0.1 2004, 13, **14**, **16**, **17**
3GPP TS 25.211 V6.1.0 2004, 17
802.11 (*See* IEEE 802.11)
802.11a (*See* IEEE 802.11a Standard)
802.11b (*See* IEEE 802.11b Standard [Wi-Fi])
802.11g (*See* IEEE 802.11g Standard)
802.11i (*See* IEEE 802.11i Standard)
802.15.3 (*See* IEEE 802.15.3 Standard)
802.15.4 (*See* IEEE 802.15.4 Standard)
802.1x (*See* IEEE 802.1x Standard)
868-MHz band, **88**, **91**
915 MHZ band, **88**, **91**

A

AAA (*See* accounting, authentication and authorization)
AAA servers with Mobile IP agents, **148**
AAAH (*See* accounting, authentication, and authorization home)
Abdelsalam, W., 202, **204**
Absolute location, 195
Abstract interaction objects (AIOs), 169
Access feedback control channel (ACH), 49, 50, 221
Access network domain, 14
Access network-serving network (AN-SN), 15
Access point (AP), 14
    802.1x address problem of rogue, 52
    connected to wired network, 41–42
    contention-free transmission, 44
    DFS operation enables, 48
    HIPERLAN/2, 47
    measures link quality on uplink channel, 48
    power management strategy, 45–46
    RC4 secret key deployed at, 55
    rouge, 52, 57
    security issues, 56–57
    uses link adaption for physical layer mode, 51
    WiseMAC, 97
Access protocol, 45
Access router (AR), 58
Access stratum, 15
Accounting, authentication and authorization (AAA), 146–149

W

Want, R., 165, 200
Warden, 157, 158
War driving. , 57
WCC (*See* wireless congestion control)
WCDMA (*See* wideband CDMA)
Wearable computing, 3–4
Weather, 84, 167, 181, 203
Weather conditions, 203, 204
Web sites, 180–181
Weiser, M., 158
Weiser, Mark, 2
WEP (*See* Wired Equivalent Privacy)
WEP2 (*See* enhanced version of WEP)
WEP and key, 56
WEP protocol, 52
Wideband CDMA (WCDMA), 7–8, 38
Wi-Fi (*See* wireless fidelity)
Willingness, 122, **123**
Window of tolerance, 157
Wired Equivalent Privacy (WEP), 52, 55–56
Wired network connected to AP, 41–42
Wireless communications
    Bluetooth radio layer, 70
    RFCOMM, 71
    seamless mobility, 175
Wireless congestion control (WCC), 49
Wireless connection admission control, 49
Wireless fidelity (Wi-Fi)
    high bit rate telecommunications, 46
    IEEE 802.11b Standard, 45–46, 63
    ISM band, 46
    networking card, 4
    power management, 74
Wireless local area network (WLAN)
    802.11e multiple backoff for traffic with
        different priorities, **54**
    attack prevention, 211
    carrier sense multiple access with collision
        avoidance (CSMA/CA), 43
    carrier sense multiple access with collision
        detection (CSMA/CD), 43
    corporate office boundary vs. WLAN
        coverage area, **58**
    CSMA/CD hidden station problem, 43
    DCF operation, **44**
    distributed coordination function (DCF),
        43
    frequency band for Bluetooth/microwave
        ovens, 67
    The hidden station problem, **43**
    HIPERLAN/2, 47–52

IEEE 802.11a Standard, 46–47
IEEE 802.11b Standard (Wi-Fi), 45–46
IEEE 802.11e Standard, 53–55
IEEE 802.11g Standard, 47
IEEE 802.11i Standard, 52–53
IEEE 802.11 Standard, 41–45
IEEE 802.1x Standard, 52
Integrating 802.11 WLAN and UMTS,
    59–62
IP over 802.11 WLAN, 58–59
location-sensing techniques, 206
Mobile IP, 146
online resources, 64
power up procedure, **62**
seamless mobility, 175–177
security Issues, 55–58
standards specified in IEEE 802.11, 63
uses of, 41
WLAN IP network, **59**
WLAN standards, a comparison of, **52**
WLAN-UMTS integration architecture, **61**
Wireless network infrastructure, 1
Wireless networks, 176
Wireless networks and services
    advance mobile phone system (AMPS), 6
    authentication, 212–213
    bandwidth, 154
    Bluetooth technology, 67
    categories of, 8
    compound wireless service, 8–9
    digital systems, 7
    domain crossing, 214
    DoS, 213
    high bit rate telecommunications, 7
    issues in mobile computing, 153
    privacy and anonymity, 213
    securing wireless devices and
        communications, 215
    vulnerability, 213–214
    wireless NIC, 56
    ZigBee Protocol, 92–94
    ZigBee versus Bluetooth, 94–95
Wireless network security
    CodeBlue Project and security, 103–104
    context-aware home and security, 80
    data transmission security, 214–216
    home area networks and security, 78
    location-sensing/location systems security,
        201, 207
    Malicious code, **219**
    mobile computing and security, 176, 181,
        186, 188
    Mobile IP and secure registration, 138–140

X

Y

Z

9 780367 388140